Carsten Maus

Toward Accessible Multilevel Modeling in Systems Biology

A Rule-based Language Concept

Logos Verlag Berlin

λογος

Bibliografische Information der Deutschen Nationalbibliothek

Die Deutsche Nationalbibliothek verzeichnet diese Publikation in der Deutschen Nationalbibliografie; detaillierte bibliografische Daten sind im Internet über http://dnb.d-nb.de abrufbar.

Zugl.: Diss., Univ. Rostock, Fakultät für Informatik und Elektrotechnik, 2013

Gutachter:

- Prof. Dr. Adelinde M. Uhrmacher, Universität Rostock

- Prof. Dr. James R. Faeder, University of Pittsburgh

- Prof. Dr. Ursula Kummer, Universität Heidelberg

Tag der mündlichen Prüfung/Verteidigung: 25. März 2013

ISBN 978-3-8325-3516-2

Logos Verlag Berlin GmbH
Comeniushof, Gubener Str. 47,
10243 Berlin
Tel.: +49 (0)30 / 42 85 10 90
Fax: +49 (0)30 / 42 85 10 92
http://www.logos-verlag.de

Acknowledgments

I thank all the people who supported me in their very own way in successfully finishing this thesis.

First and foremost, my sincerest thanks go to my supervisor Lin Uhrmacher for her endless confidence in me and my research. She had always an open door for me and taught me a lot, not just about modeling and simulation. It was a great pleasure for me to work with her. I am also grateful to Jim Faeder, who helped me to expand my knowledge about rule-based modeling during a very pleasant two-month stay at his lab in Pittsburgh, and to Ursula Kummer, who accepted to review my thesis only a few weeks in beforehand.

I also want to thank all my colleagues of the modeling and simulation group and the interdisciplinary research training school dIEM oSiRiS for their support and the great time we have spent together in Rostock. Special thanks go to Mathias John for fruitful discussions on formal modeling languages, Stefan Rybacki for implementing many of my ideas, as well as to Jan Himmelspach and Roland Ewald, who both supported me in countless situations and made it easier for me to start living in a new town.

Lastly, but not less importantly, I thank my parents for always believing in me, and in particular my lovely wife for her endless patience and for keeping me free of ties in the last stage of writing. This thesis could have not been finished in time without you.

Contents

List of Figures

List of Tables

Chapter 1

Introduction

Systems biology aims at analyzing the complex interrelationships within biological systems in order to understand their functional behavior (Kitano, 2002b). Though often claimed to rely on a fairly new methodology (Ideker et al., 2001), perceiving life in terms of complex systemic interactions among various components has in fact a long tradition (Wolkenhauer, 2001), dating back to rather abstract ideas of ancient philosophers and – more specifically – to concepts developed by the founders of general systems theory like Norbert Wiener (1948) and Ludwig von Bertalanffy (1968). Anyways, promoted by the recent progress in obtaining large amounts of high-quality data, systems biology has nowadays become a well-established field of research and can be considered to be *the* main scientific paradigm in modern life sciences for studying complex biological systems at various organizational levels, such as genetic, molecular, cellular, and even at the level of whole organisms (Butcher et al., 2004). Thereby, research projects often run through an iterative workflow process (Kitano, 2002b), starting with experimental data acquisition in the "wet-lab", followed by data analysis and the development of hypotheses how the experimental observations could be explained. To deal with the complexity of biological systems, these hypotheses are in a next step typically transformed into formal representations (models) allowing to be analyzed computationally, e.g., by performing simulation studies. Results of these "dry" experiments can thereafter be used to design new "wet-lab" experiments, by which the cycle becomes completed and may start again from the beginning.

The research described in this thesis concerns the *modeling* part of this systems biology workflow, i.e., the transformation of a certain hypothesis from an informal *mental* into a computationally analyzable *formal* model representation. The focus thereby lies on the *accessibility* and *expressive power* of modeling methods for describing so called *multilevel models*, aiming at the integration of processes at multiple levels of organization.

1.1 Motivation and Problem Statement

The process of modeling always means to abstract from what we consider to be the real world, or as Olaf Wolkenhauer states: *"[. . .] understanding arises from transforming, abstracting one reality into another–through modeling."* (Casini, 2011, p. 143). Finding the right abstraction level, i.e., deciding on which pieces of information shall be part of the model, denotes a challenge of its own and has – due to its creative aspects – much in common with art. Once *"having determined what kind of knowledge will be contained in the model, the next task is to choose how best to represent it"* (Leitch et al., 1999, p. 436f), which, however, is no trivial task either in many cases. The difficulty lies in a trade-off between two contradictory goals:

1. Finding an appropriate representation format that allows for expressing all relevant information.

2. Keeping the formal model description as simple as possible, e.g., to minimize the chance of introducing errors and to communicate hypotheses and results to people that are not familiar with all details.

Other aspects like simulation performance issues may play an additional role when deciding for a particular model representation.

A plethora of different modeling formalisms has emerged during the past decades, many of which have been successfully applied to modeling diverse biological systems (de Jong, 2002; Machado et al., 2011). Also, various user-friendly software tools have been developed to aid modelers in their task and thereby increase the accessibility of modeling, e.g., COPASI (Hoops et al.,

2006) and Snoopy (Heiner et al., 2008b). Thereby, most of the existing formalisms and tools focus on describing dynamic systems at a single organizational level, which is not surprising, as according Occam's razor modeling projects typically start simple – in particular if only little is known about the system of interest – and yet the relationships between components at a single level may become rather complex. Thus, in the past there has been often simply no need for alternative model representations.

In recent years, however, accumulated knowledge and increasing amounts of available data progressively indicate that examining a single level in isolation may not suffice to explain certain observations (Noble, 2006). Therefore, a shift of the needs toward so called *multilevel* or *multiscale* modeling techniques can be observed (Degenring et al., 2003, 2004; Uhrmacher et al., 2005; Derosa and Cagin, 2010), to support the description of interrelated dynamic processes across different levels of organization, such as a multicellular system that consists not only of multiple interacting cells, but in which each individual cell has also its own internal dynamics and processes at both the subcellular as well as cellular level may influence each other. Increasing importance of multilevel modeling in systems biology is particularly demonstrated by the quantity of according models that have recently been published. Also special sessions at major systems biology conferences like the annual *International Conference on Systems Biology* (ICSB) indicate a growing interest in this topic. However, conventional formal modeling approaches applied in systems biology are typically not well-suited for describing such models and thus usually alternative modeling methods are employed.

So far, many multilevel models as well as respective simulation algorithms are implemented more or less from scratch by directly using higher programming languages like C or Java (e.g., Ribba et al., 2006; van Leeuwen et al., 2009; Dexter et al., 2009). While being extremely flexible, this modeling approach is error-prone, time-consuming, requires profound knowledge in software programming, and can be generally considered to be inaccessible to most biologists. Moreover, such models are typically strongly intertwined with their simulators as well as other code required for performing simulation experiments – e.g., to observe simulation results – and therefore do not comply with the principle of separating concerns in modeling and simulation (Ewald et al., 2010).

Others employ agent-based modeling toolkits like MASON (Luke et al., 2005), NetLogo (Tisue and Wilensky, 2004), or FLAME (Holcombe et al., 2012), e.g., Zhang et al. (2007) and Sun et al. (2009), or special-purpose multiscale modeling and simulation frameworks like CompuCell3D (Cickovski et al., 2007), Simmune (Angermann et al., 2012), OpenAlea (Pradal et al., 2008), or CellSys (Hoehme and Drasdo, 2010) to encode their models. These approaches, however, are often rather rigid in a sense that they restrict users to describing particular kinds of phenomena and model structures. For example, to represent spatial dynamics at the tissue level, the CompuCell3D framework implements a parameterizable cellular Potts model, which is basically a generalized cellular automaton; describing other kinds of models is not supported. In general, even if the structure and dynamics of models can be modeled more freely, such tools typically lack a universal language for describing dynamic processes at each level of the model similarly. Instead, dynamics at different levels need to be formalized differently by using different editors or different syntactical elements of a model specification language. In this context also frequently found, advanced programming skills may still be required to combine or integrate different levels, although the actual dynamics at each level might be formulated in a more simple way. Other approaches, which adapt general hierarchical modeling formalisms for multilevel modeling in systems biology, like Degenring et al. (2003, 2004), suffer from lacking explicit support for interrelating the different levels and from conveniently representing biochemical reactions.

Taken all together, the aforementioned approaches for multilevel modeling of biological systems have major limitations either (or even both) with respect to *expressivity*, i.e., the kind of systems and processes that can be described, or with respect to *accessibility*, i.e., the ease of use and succinctness of representing multilevel models formally.

The aim of this thesis is therefore to contribute new insights and concepts to the field of modeling methodology and thereby facilitate multilevel modeling in systems biology.

1.2 Contribution

According to the aim of this thesis to facilitate multilevel modeling of biological systems, the work concentrates on the following two major aspects:

1. Comparing diverse existing modeling approaches with respect to their suitability for describing biological systems at multiple levels and thereby identifying useful constructs and features for this task.

2. Based on these findings, developing a tailor-made language concept for accessible multilevel modeling in systems biology.

Therefore, at first a simple toy model is defined serving as a running example throughout this thesis and denoting the basis for a comparison of different modeling approaches. The example model describes a simple Notch signaling system (Baron et al., 2002; Schweisguth, 2004) at different organizational levels. It consists of multiple proliferating and interacting cells, ordered in a one-dimensional spatial neighborhood. Each cell thereby consists of different subcellular compartments, which in turn contain different molecular species, i.e., Notch receptors and Delta ligands. At the molecular level, dynamic processes are dedicated to the turnover of Notch and Delta, their translocation from one compartment into another, as well as interactions between both species, constrained to only take place if Notch and Delta are located within different but adjacent cells. Processes at higher organizational levels are dynamically increasing compartment volumes and the traversal through different cell cycle phases, eventually resulting in the division of the cell, by which the total number of cells and their spatial neighborhood is changing. Causation between different organizational levels includes the effect of growing compartment volumes on the rate of bimolecular reactions taking place within these compartments (downward causation) as well as cell cycle phase transitions in dependence on the amounts of intracellular proteins (upward causation).

In the next step, this informal multilevel model is transformed – as far as possible – into several formal model representations. Thereby, modeling formalisms with a clearly defined syntax and semantics are considered only, among those diverse *flat* formalisms, such as ODEs, Petri nets (Petri and

5

Reisig, 2008), and the π-calculus (Milner, 1999), but also modeling languages that have an explicit notion of *hierarchy*, such as BioAmbients (Regev et al., 2004), DEVS (Zeigler, 1976), and Bigraphs (Milner, 2009). On the basis of the exemplary model encodings, general properties of these different formalisms and their suitability for multilevel modeling of biological systems are thoroughly discussed. Diverse aspects are thereby taken into account, such as a formalism's support for describing dynamic model structures and how upward and downward causation can be realized. This extensive examination of different modeling formalisms is partially based on the following publications:

> Ewald, R., Maus, C., Rolfs, A., and Uhrmacher, A. (2007). Discrete event modelling and simulation in systems biology. *Journal of Simulation*, 1(2):81–96.

> Maus, C., John, M., Röhl, M., and Uhrmacher, A.M. (2008). Hierarchical Modeling for Computational Biology. In Bernardo, M., Degano, P., and Zavattaro, G., editors, *Formal Methods for Computational Systems Biology*, volume 5016 of *LNCS*, pages 81–124. Springer.

As one main result of this comparative study it turns out that attributed model entities and flexible reaction constraints are not only essential for modeling spatial cell biological processes in the π-calculus, as has been shown by John (2010), but they are also key features for representing dynamic behavior at different organizational levels as well as describing interlevel causation. In addition, the support for explicit representations of hierarchically nested model structures denotes an essential aspect to diminish redundancy and particularly to capture important characteristics of a multilevel system without further explanation. Other valuable language features refer to describing dynamically changing structures (Uhrmacher and Zeigler, 1996; Uhrmacher, 2001) and the compositionality of model descriptions. Last but not least, diverse examples indicate that reaction-centric modeling paradigms – as employed, e.g., by rule-based approaches (Hlavacek et al., 2006) and Petri nets – seem to be more accessible and generally more practical for describing biological systems at multiple levels than object-centered approaches, like DEVS or the π-calculus.

This is mainly due to the straightforwardness of describing biochemical processes, but also taking certain side-effects into account to model interlevel causation can typically more straightforwardly and succinctly be realized by using a reactions metaphor.

Based on the above findings, ML-Rules, a general-purpose modeling concept for describing biological models at multiple interrelated levels is presented. The developed concept employs a novel rule-based modeling approach, where each entity (called species) may contain a set of further entities to represent hierarchically nested structures. Rule schemata help reducing the size of models and equally important, add the required flexibility to express dynamics at different levels in a general manner. In addition, species may have assigned attributes to describe different states of species at any level, including those denoting the containers of hierarchical model structures, e.g., to represent membrane-bound compartments with dynamically changing properties like an increasing volume. Arbitrary rate laws and reaction constraints allow for flexibly describing upward and downward causation as well as certain behavioral abstractions like approximated enzyme kinetics. Moreover, describing simple spatial dynamics beyond compartmentalized structures becomes also possible. The discussed approach has been published in the article below.[1]

> Maus, C., Rybacki, S., and Uhrmacher, A.M. (2011). Rule-based multi-level modeling of cell biological systems. *BMC Systems Biology*, 5:166.

By introducing additional syntactic layers, the accessibility of ML-Rules may be further increased compared to its basic syntax. A graphical representation inspired by Milner's Bigraphical Reactive Systems (Milner, 2001) may increase the readability of rules consisting of nested species patterns. Independent from a graphical or textual notation, identifying species attributes by name instead of position may additionally improve the readability. Moreover, this allows for a *"don't care, don't write"* approach, by which rules may become more succinct and their generality can be increased. The same holds true for

---

[1]Oury and Plotkin (2011) have published a rather similar approach at nearly the same time. Differences between ML-Rules and their approach are briefly discussed in the conclusions at the end of this thesis.

generic species, where wildcards instead of defined species names are used in order to reduce the number of rules needed. Another meaningful extension of the basic ML-Rules concept is the introduction of functions on solutions (multisets of species), in particular to count certain species being part of a solution and to decompose a given solution into multiple sub-solutions, e.g., to describe processes like cell division. Unlike the previously described extensions, functions on solutions cannot be introduced as purely syntactic sugar. However, they do not conflict with the basic concepts of ML-Rules and can thus straightforwardly be incorporated.

The final contribution of this thesis is a demonstration of the usefulness of the presented language concept for multilevel modeling of biological systems. Therefore, at first the encoding of the running Notch signaling example is being discussed. The second case study describes a model of fission yeast cell proliferation at multiple interrelated levels of organization, from a subcellular biochemical control circuit up to a spatially discretized environment in which diffusible pheromone molecules and cells may move in space. This second example is published as part of the aforementioned article in BMC Systems Biology, but has been slightly adapted for this thesis.

1.3 Structure of the Thesis

The structure of this thesis follows the previously outlined contributions. Hence, after introducing some general principles and concepts of hierarchical multilevel modeling in Chapter 2, the subsequent Chapter 3 provides an informal description of the recurring Notch signaling model. Thereafter, Chapters 4 and 5 discuss the applicability of diverse flat respectively hierarchical modeling approaches for describing multilevel models in systems biology. Chapter 6 presents the basic rule-based multilevel modeling concept of ML-Rules, which becomes extended and syntactically refined in Chapter 7. Two case studies in Chapter 8 demonstrate the usefulness of the presented approach. In the final Chapter 9, a short summary of this thesis is given and worthwhile directions for future work are discussed.

Chapter 2

Multilevel Systems Modeling

In this chapter, first a general introduction to complex systems and levels will be given. This part is intended to be as general as possible. Thus, to illustrate the theories, examples will be presented that do not exclusively belong to the biological realm. Important basic concepts of multilevel modeling are briefly discussed in Chapter 2.2. Finally, by relating these concepts to the concrete application of systems biology, Chapter 2.3 underlines the importance of multilevel modeling approaches in this particular field of research.

2.1 Systems, Complexity and Multiple Levels

2.1.1 Emergence of Complex Systems Behavior

Definition *"A system is a set of interacting units with relationships among them."* (Miller, 1978, p. 16). Together, they form an integrated ensemble that is somehow separated and distinguishable from its environment.

In reality, most systems are complex, and they are ubiquitous. A living organism is a complex system, just like climatic, ecological, economical, or social systems are. Although complex systems typically coincide with a large number of involved components (also called elements, objects, units, entities or parts), complexity is only improperly defined by the sheer size of a system. The dynamic interactions among its components makes a system being complex. In this context, related and commonly found keywords are, for instance, nonlinearity, dynamic networks, pattern formation, adaptation, and self-organization. However, an exact and universal definition of complexity does not exist.

A universal but rather unspecific feature is the manifestation of system properties that are not obvious from studying the component properties, or to say it in the words of Gallagher and Appenzeller (1999), a complex system is *"one whose properties are not fully explained by an understanding of its component parts"*. Complex system states and dynamics *arise* or *emerge* out of the rather incomplex interactions among the system's components (see, e.g., Weaver, 1948; Polanyi, 1968; Anderson, 1972). Consequently, the illustrative term "emergence" has become established for such phenomena.

Early concepts of emergence date back to descriptions made by the Greek philosopher Aristotle (M'Mahon, 1857, book VIII), commonly reproduced by the popular simplified quotation: *"The whole is greater than the sum of its parts."* For instance, the observed phenomenon of liquid water at room temperature is an emergent property of the whole (the system 'water droplet'), while its parts or components (the individual H_2O molecules) do not show liquid properties on their own nor can the emergent property be easily derived from the physicochemical interactions between those. Another example of emergence is the blood pumping heart. None of its various components (like muscle and valve cells) shows properties of a pump itself, but the complex and concerted interplay between electrical stimuli, biochemical interactions and physical forces leads to the emergent behavior of blood transportation due to regular contractions.

Two different notions of emergence can be distinguished from each other, typically designated to be *weak* or *strong*. Weak emergent properties are – at least in principle – predictable from the behavior of system components. It does not imply that a prediction is easy to achieve nor that it is practically attainable anyhow. However, weak emergence goes along with the ideas of reductionism, i.e., any behavior can be deduced from fundamental behavior of the smallest constituents of a system. In contrast, an emergent phenomenon is considered to be strong if it is irreducible as a matter of principle, not just in practice. It thus denotes a totally new essence and peculiarity. As complex system behaviors easily seem to be irreducible, although in fact they may not (as they may be the result of weak emergence), the ongoing controversy between those who refuse the existence of strong emergence (e.g., Bedau, 1997) and those who advocate for it (e.g., Chalmers, 2006; Kauffman, 2008), is a rather philosophical debate and therefore of only limited relevance for this thesis.

2.1.2 Structural Organization

Resulting from emergence, another *"key property of complex systems is their self-structuring in conditioning levels"* (Iordache, 2011, p. 1), such that a hierarchy of levels is formed (see also Simon, 1962; Salthe, 1989; Salthe and Matsuno, 1995). Thereby, each level may consist of different components and interaction laws. Moreover, as levels emerge from complex interactions within systems, entities at one level appear to be formed by interactions among entirely different entities, i.e., by individual complex systems at another level. Hence, the world (and probably everything beyond) seems to be made of systems of systems.

Let us consider an illustrative example: Animals in a complex ecosystem are interacting with each other according to certain laws, e.g., a food chain in which wolves feed on sheep with a certain probability. However, the individual animals seem to be also complex systems on their own, where various components like organs, bones and other structures are interacting with each other, resulting in emergent phenomena that are characteristic for a sheep or a wolf respectively. Each organ of each individual animal, in turn, looks like the emergent property of billions of interacting cells, each cell like the product of interacting molecules, and so forth.

In the middle of the last century, scientists like Paul A. Weiss (1969, 1971) and Ludwig von Bertalanffy (1950) identified the ubiquity of such hierarchies within complex systems. Thus, the principle of hierarchical organization became a fundamental aspect of the general systems theory (von Bertalanffy, 1968).

In an abstract sense, a hierarchy describes an arrangement of certain elements, such that one element is *below* or *above* another one (see, e.g., von

Bertalanffy, 1950; Webster, 1979; Salthe, 1985). To be more concrete, an organizational hierarchy is often formally defined by a partially ordered set, i.e., a set of elements ordered by binary relations that are reflexive, antisymmetric and transitive (cf. Bunge, 1969; Simon, 1973). However, while such ordered sets provide clear arrangements of distinct elements, the above definition alone is insufficient for describing the complex structural organizations of multilevel systems, as it lacks any qualitative property of relations. Not explicitly addressing, this simple and rather universal definition says little about what the actual hierarchical *levels* are. Hence, further aspects need to be considered.

Let us make a small step back for that. In hierarchy theory, it is widely acknowledged that hierarchical organization is ubiquitous but primarily a conceptual construct of human cognition: *"Whether Nature is truly organized hierarchically is moot. Man's perception of nature is hierarchical."* (Webster, 1979, p. 120). Timpf (1999, p. 128) points out that *"'hierarchization' is one of the major conceptual mechanism to model the world."*

Two different general forms of hierarchies are known: classification (also taxonomic or subsumptive) and composition hierarchies (see, e.g., Grene, 1969; Uhrmacher, 1992; Zylstra, 1992; Pumain, 2006; Maus et al., 2008). The synonymous terms "specification hierarchies" and "scalar hierarchies" are used by Stanley N. Salthe (1991, 1993). He argues that *"any natural system can be analyzed from the point of view of the scalar hierarchy [...] or from that of the specification hierarchy"* (Salthe, 1993, p. 93), meaning that hierarchies are conceptual tools for analyzing systems rather than compulsory actualities. From this point of view, we are free to chose the most appropriate one. However, both general hierarchy concepts follow quite different perceptions.

A classification or specification hierarchy follows the concept of categorization and specification (Lakoff, 1987). Its focus lies on marking the *"qualitative differences of different realms of being"* (Salthe, 2001). Different levels are determined by different descriptive views, and their ordering emerges from refinement and abstraction relations among these different views. Hence, a classification hierarchy is ontological and the ordering of elements follows a path from the general to the most highly specific description (see also the section on abstraction levels on page 23ff). A famous example is the taxon-

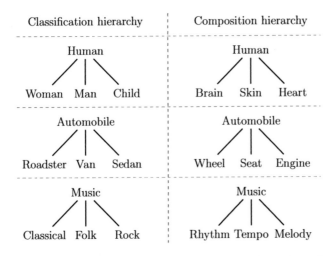

Figure 2.1: Comparison of classification and composition hierarchies. In a classification hierarchy, the superior denotes a general class of its subordinate types or instances. A composition hierarchy relates elements according to wholes comprising parts.

omy of biological classification with its numerous taxonomic ranks: domain → kingdom → ⋯ → genus → species.

In a classification hierarchy, the relationship between a superordinate element and its subordinates is that of a generic class of something and specific types or instances of it. For example, both mouse and whale are different types of the class called mammal. Subordinates inherit the properties of their superordinate, e.g., a whale has characteristic properties of a mammal. At one level, all elements are thus of the same natural kind and each lower level in the hierarchy denotes a refinement of its superior level. Accordingly, relations between different elements (from lower to higher level) follow an "is-a" scheme: a mouse *is a* mammal and a whale *is a* mammal too. Although biological taxonomies today typically rely on cladistic methods such as phylogenetic classifications based on genome sequence similarities, a classification or specification hierarchy often *"labels itself as human discourse, an obvious social construction"* (Salthe, 1993, p. 93).

In comparison, an alternative composition or scalar hierarchy *"is often taken to be a necessary, 'objective' datum"* (Salthe, 1993, p. 93). Such hierarchy *"is one of parts nested within wholes"* (Salthe, 2001), i.e., it relates elements according to the parts that build up a system. For example, organs like the brain, skin, and heart are parts of a human body. That means, the ordering of elements of a composition is determined by "is-part-of" relations (or "consists-of" when starting from the superior) among different components (Webster, 1979; Uhrmacher, 1992). Consequently, when dealing with physical objects, a compositional hierarchy implies an ordering of levels according to spatial scales, i.e., the composition of elements is constrained in a way such that small parts may be enclosed by larger ones only. This kind of composition is called "nesting" and implies that each element has one superordinate at most. However, a compositional hierarchy may comprise also nonmaterial components like a song that consists of vocals and instrumental sounds. In this case, a composition may be also of a non-nested form, where elements may be part of more than one superordinate, e.g., a person that is a member of different equally ranked organizations.

Figure 2.1 provides some further examples and illustrates the disparity between composition and classification hierarchies. If not specified differently, hierarchies in this thesis are of a compositional form and defined as follows:

Definition A hierarchy is a nested set of components in which subordinates are being part of their superior. It complies mathematically with an acyclic directed and rooted tree in graph theory, i.e., a partial ordering that is anti-symmetric, reflexive, and transitive. A hierarchical level is a set of components that share the same distance to the hierarch, i.e., the root node of the tree.

The last sentence of the above definition needs maybe some further explanation: As has been described by Simon (1973), a partial ordering allows for hierarchies not only in the sense of Matryoshka dolls and Chinese boxes, where each level consists of only one further box. Instead, the above definition of hierarchy and level allows to arbitrarily enclose multiple subsets, i.e., to build Chinese boxes where each box may enclose a whole set of multiple small boxes, each may consist of another set, and so forth. This in turn implies that a level is defined by its distance to the highest element of the hierarchy and

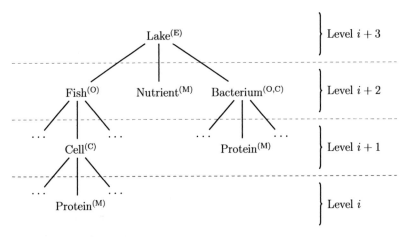

Figure 2.2: Hierarchical organization of an ecosystem, i.e., a lake, in which the same level may consist of elements of different kinds. Legend: (E) ecosystem, (O) organism, (C) cell, (M) molecule.

that certain elements at the same level may be part of different subsets of the hierarchy. Thereby, different subsets (subtrees) are separated from each other. A tree-like hierarchy thus constitutes *"both a vertical separation that isolates each level from levels above and below, and a horizontal separation that segregates the components of any level into groups"* (Webster, 1979, p. 122). Unlike in classification hierarchies, a single level may then also consist of elements of totally different kind.

Let us consider an example of an ecosystem organized into different levels, as shown in Figure 2.2. While in this example higher organisms like fishes are hierarchically organized into cells and molecules (e.g., proteins), simple unicellular organisms like bacteria lack the intermediate cellular level – or to be more precisely, the organism level and the cellular level of a bacterium coincide. Also, molecules are not solely found within cells, but are also contained by the ecosystem directly, i.e., nutrients (or water molecules) within the lake. Consequently, the arrangement (or ordering) of levels may depend on the nested system structure, such that, for instance, bacterial proteins act on the same level as cells within a fish.

2.1.3 Inter-Level Causalities

Causality (or causation) is commonly understood as the relation between two events, where one event (an effect) is the consequence of another one (the cause). Similarly to the question whether nature is truly organized hierarchically (cf. previous section), the philosophical discussion whether causality does really exist (see, e.g., Hume, 1739; Bunge, 1959; Schlegel, 1961) is considered to be rather irrelevant for this thesis, as nature and hence the models of our world are understood by laws of causation. Therefore, the question is not *whether* but *how* different levels are causally interrelated.

As already mentioned, properties of different levels emerge from complex system dynamics. Webster (1979, p. 127) remarks: *"Behaviour at any level is explained in terms of the level below, and its significance is found in the level above."*. Therefore, *"in any organisation the higher level depends upon the lower"* (Feibleman, 1954, p. 60), i.e., complex interactions among components of lower levels tend to result in emergent properties that manifest themselves as higher levels of organization. Dynamics at these higher levels may then cause once superior level properties, and so forth. Consequently, a strong bottom-up causal chain can be observed within hierarchically organized systems. For example, molecules interact with each other according to certain laws. Thereby, the organizational structure and characteristic bahavior of a functional living cell may emerge. Many of those cells may now interact with each other and thereby cause the emergent property of an organ and finally of higher organisms. Hence, complex hierarchically structured systems involve "upward causation".

Definition Upward causation describes a process by which parts of a system influence states and behavior of the system as a whole, i.e., states and behavior at higher levels of organization.

Upward causation complies with the typical reductionistic systems view, where the whole is reducible to and fully determined by the behavior of the small parts of lower levels. That is why hierarchy definitions that are entailing unidirectional domination of higher levels over lower ones are problematic for describing hierarchies in nature, as has been observed by Bunge (1969). However, if the whole is controlled by small parts, who will control the controllers, or *"sed quis custodient ipsos custodes?"* (Glanville, 1990). The prevalent existence of upward causation does not imply absolute nonexistence of conversely directed causality, i.e., causation in a top-down manner. According to Donald T. Campbell (1974, p. 180ff), quite the opposite is true: *"all processes at the lower level of a hierarchy are restrained by and act in conformity to the laws of the higher level"* and *"the higher level laws are necessary for a complete specification of phenomena both at that higher level and also for lower levels"*. Hence, besides upward causation, complex hierarchically organized systems also include "downward causation".

Definition Downward causation is a process by which the system as a whole influences states and behavior of its parts, i.e., states and behavior at lower levels of organization.

For example, think over the situation of an animal which is suddenly overtaken by death due to an external cause, e.g., due to drowning in a river or getting eaten by a predator. Such high-level processes have serious consequences for the whole organism and many of its levels below. Although the sudden death does not result from processes at lower levels, many of the enclosed structures (like organs, tissues, cells, etc.) will get disrupted as a consequence. Even low-level molecules like proteins and lipids will finally get digested and thereby broken down into smaller parts. The system as a whole (ecosystem) strongly influences how its parts (the animal and its enclosed components at lower levels) behave.

So, in addition to intra-level causality, complex behavior of hierarchical systems is also determined by both kinds of inter-level causality, i.e., causation in an upward as well as downward direction (Corning, 2002). All processes at all levels are more or less interrelated (Simon, 1962), by which *"the whole is to some degree constrained by the parts (upward causation), but at the same time the parts are to some degree constrained by the whole (downward causation)"* (Heylighen, 1995). Therefore, *"the GS [i.e., general systems] approach, and the theories prompted by it, are neither atomistic nor holistic"* (Bunge, 1977, p. 89). In other words, it is the duality of upward causation and downward causation that makes the systems approach being different from the antonymous ideas of pure reductionism or holism.

2.2 Multilevel Modeling – Brief Overview of Related Terms and Concepts

In the previous section, a brief introduction to complex systems and their structuring into distinct levels has been presented. Multilevel modeling deals with specifying system models at multiple levels, i.e., in such models different levels can be distinguished. However, before we start going into detail, some further essential terms and general concepts need to be discussed.

2.2.1 Modeling and Simulation

When speaking about "modeling", the signification of "model" should be clarified at first, as *modeling* describes the process of *model building*. In the most abstract and general meaning, a model is some kind of representation or image of something else. A model's objective is to answer questions about what it is representing: *"To an observer B, an object A* is a model of an object A to the extent that B can use A* to answer questions that interest him about A."* (Minsky, 1965). Another definition – according to Cellier (1991) also attributed to Minsky – emphasizes the relationships between system, model, and experiment:

Definition *"A model (M) for a system (S) and an experiment (E) is anything to which E can be applied in order to answer questions about S."* (Cellier, 1991, p. 5).

Please note, the above definition is highly universal, i.e., a model does not necessarily require to be formally specified and also the term "experiment" remains rather vague. It could be, for instance, the exerted thinking about a completely mental or informal pen-and-paper model. An experiment performed on a model is called "simulation" (Cellier, 1991).

Definition *"Simulation is experimentation with models."* (Korn and Wait, 1978).

Computational simulation implies that the model must be somehow formally specified, so that a certain simulation algorithm can be performed on it. A formal model specification is typically realized either in the form of mathematical equations, in a special-purpose modeling language, or encoded in a highly versatile programming language.

In addition, for reliable modeling and simulation studies, a *separation of concerns* is desired, so that, for instance, a modeler can concentrate on his or her primary concern of building a model (see Zeigler et al., 2000). A model specification that is independent from a concrete implementation of the simulation algorithm, i.e., the simulator, also allows to study the model from different perspectives, e.g., by defining a modeling language with alternative semantics for the same syntax. Lastly, separation of concerns may enhance trust in the validity of certain experiments, as one may interchange different independently developed models and simulators and compare the results.

2.2.2 Multilevel vs. Multiscale

Discussions on hierarchical systems modeling sometimes become quite complicated or confusing due to varied naming of respective methodologies. In particular, "multilevel" and "multiscale" are oftentimes synonymously used (see, e.g., E and Engquist, 2003; van der Hoef et al., 2006; Stamatakos, 2010). However, although often describing the same concepts, in recent years a rather distinct usage of both terms can be observed.

Methods for modeling and simulation of hierarchical systems are typically denoted "multiscale" (e.g., Fish, 2009; Engquist et al., 2009; Derosa and Cagin, 2010; E, 2011; Dada and Mendes, 2011), whereas the term "multilevel" is more frequently used to describe statistical analysis methods like covariance analysis and linear regression models applied to multilevel data, i.e., empirical data that have been obtained from different organizational levels like school grades at the levels of classes, schools, and districts (see, e.g., Kreft and de Leeuw, 1998; Snijders and Bosker, 1999). On the other hand, also oftentimes the boundaries between such classifications and methods are blurred and thus a clear and common definition is missing (see also Bittig and Uhrmacher, 2010).

Although the motivation mostly seems to originate in the study of interlevel correlations, using the *multiscale* terminology for certain approaches might became rather popular as it inherently reflects the fact that different levels often coincide with different temporal and spatial scales. For instance, while the diameter of a typical bio-molecule is about 10^{-9} m and the according time scales of molecular processes range from nanoseconds to milliseconds, scales are orders of magnitudes higher when looking at the organism or ecosystem level (meters and years). For the same reason, Salthe introduced the term "scalar hierarchy" to define nested compositions: *"Entities existing at different scalar levels have characteristically different time scales"* (Salthe, 1991, p. 252) and in a scalar hierarchy, *"classes are rankings by scale, so that the more inclusive classes have as members objects of larger scale"* (Salthe, 1993, p. 93).

The integration of dynamic processes that are operating at highly diverging spatiotemporal scales requires special attention with respect to simulation methods (Dada and Mendes, 2011). Otherwise, computational resources like simulation time and memory may easily exceed reasonable amounts. The reason for that can be best made clear with an illustrative example. For instance, compared to the lifespan of an entire living organism, molecular events taking place at lower levels of organization are several orders of magnitudes faster. Thus, although low-level states are changing frequently, from the point of view of a high-level observer the dynamic molecular system may appear to be in a quasi-steady state, i.e., the concentrations of molecules may seem to be constant. Similarly, what is a water droplet for a human is like a whole universe for a unicellular organism. Hence, for a low-level observer, dynamics at a higher level may be out of scope. Simulating a model comprising rates of different scale entails the same problems. For example, in a certain range of time, an enormous number of low-level events may happen, while nothing is changing at higher levels due to significantly slower dynamic processes. Therefore, special simulation methods like stiff system equation solvers (Cohen and Hindmarsh, 1996; Shampine and Thompson, 2007) or hybrid simulation approaches (Takahashi et al., 2004; MacNamara et al., 2008) may be needed.

Even though dealing with modeling nested compositions in the sense of *scalar* hierarchies (cf. Chapter 2.1.2), in this thesis "multi*level* modeling" is the terminology of choice as it concentrates on modeling the hierarchical *level* structure of biological systems rather than on simulation issues arising from the combination of highly varying temporal and spatial scales. Also, although multilevel modeling often implies to describe dynamics at different scales, one can definitely think of hierarchically nested systems where processes at different levels do not significantly differ in space and time, e.g., molecular processes within different nested compartments of a cell, like the nucleus, cytosol, or vesicles (see also Chapter 2.3.1). Thus, "multiscale" should be reserved for naming approaches in which linking different scales is the subject of interest.

2.2.3 Hierarchical Model Structures

Even when agreeing on a basic vocabulary with respect to levels and scales, a consensus about a general definition of "multilevel modeling" does not exist. The question is, what makes a *model* being multilevel?

Although describing entities and dynamic processes at multiple levels is typically considered to be an essential characteristic of multilevel modeling, in some work, the distinction between different levels becomes not necessarily apparent from the model description. For example, Lai et al. (2009) present a multilevel model of subcellular and cellular processes by using a set of differential equations, i.e., certain equations describe the subcellular dynamics while others describe the dynamics at the cellular level. In this case, the structure of the model description does not differ from a single-level model in principle and thus perceiving different levels is difficult without profound knowledge about the modeled system. The multilevelness is described implicitly only (see Chapter 4 for more details and illustrative examples).

Others highlight the importance of hierarchical model structures as an essential and salient feature of multilevel models: *"composition and interaction determine the overall structure of a model in general and of a multi-level model in particular."* (Uhrmacher et al., 2005, p. 80). A distinction between different hierarchical levels can be achieved through specific model elements supported by according modeling languages (Zurcher and Randell, 1968; Mesarović et al.,

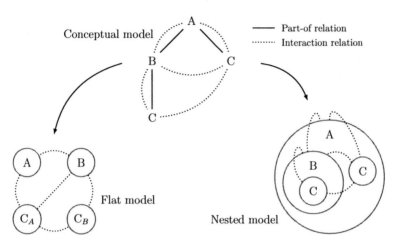

Figure 2.3: Difference between flat and nested model structure. In a flat model, all elements reside equally side by side and thus the model is more or less unstructured. Hierarchical relations of the conceptual model can be described implicitly only through according annotations. A nested model structure makes hierarchical relations explicit, i.e., both A and B enclose one entity C, while B is part of A.

1970; Zeigler, 1984). In comparison to "flat" (non-hierarchical) approaches, a hierarchical modeling approach includes means for describing hierarchical relations between model entities, e.g., by embedding certain elements within others. This allows for translating the actual hierarchy of a conceptual model in an explicit way (see Figure 2.3 and Chapter 5). Thereby, understanding as well as accessibility of models may increase (Luna, 1993; Chwif et al., 2000), as crucial knowledge about a modeled system may be mimicked more appropriately. Also, by following a *divide and conquer* strategy, modeling large systems may be facilitated: *"Partitioning models into hierarchical subcomponents can be crucial for the manageability of complex models."* (Daum and Sargent, 1999, p. 1471).

A prominent example of a hierarchical modeling formalism is the Discrete Event System Specification (DEVS) by Zeigler (1976, 1984). In DEVS, model components may be freely placed into coupled models to form nested compositions, thereby allowing for the specification of generalized hierarchically structured models. Model components can be seen as timed automata, which might be connected with each other for interaction via communication interfaces (for more details, see Zeigler et al. (2000) and Chapter 5.2.2).

Along with hierarchical structuring, a frequently found concept is object-orientation (see, e.g., Zeigler, 1987; Uhrmacher, 1995; Takahara and Shiba, 1996; Möhring, 1996). Different (parameterized) instances of a rather generally specified component class may be created. Thereby, model complexity in terms of size, lines of code, or number of components, may be significantly reduced compared to model specifications without such object instantiation. Zeigler's System Entity Structure (SES) (Zeigler, 1984) is a formal tool for describing the structure of hierarchically nested models in an object-oriented manner. An SES includes characteristics of both, classification as well as composition hierarchies, as it is a *"structural knowledge representation scheme that systematically organizes a family of possible structures of a system. Such a family characterizes decomposition, coupling, and taxonomic relationships among entities."* (Zeigler and Sarjoughian, 1999). In combination with component repositories (e.g., consisting of DEVS models), an SES can be used to automatically generate concrete hierarchical model instances.

Last not least, another important aspect in hierarchical systems modeling is the dynamic change of model structures (Zeigler and Praehofer, 1990; Uhrmacher, 1993; Uhrmacher and Zeigler, 1996; Uhrmacher, 2001). Dynamic variable model structures are particularly relevant for reflecting compositional change of hierarchies, e.g., by adding or removing certain entities, but also by moving entities to other levels. In Chapter 5, different concepts of variable structure modeling are discussed in more detail.

2.2.4 Multiple Abstraction Levels

Tightly coupled to hierarchical systems modeling is the integrated description of processes at different levels of *abstraction*.

In general, *"an abstraction of some system is a model of that system in which certain details are deliberately omitted"* (Smith and Smith, 1977, p. 105), and *"the level of abstraction of a model determines the amount of information that is contained in the model."* (Benjamin et al., 1998, p. 391). Abstraction is a relative measure and its level decreases with a larger quantity of information, i.e., a low-level abstraction shows more details than a high-level abstraction (Timpf, 1999). Originated and applied in various fields of application, a multitude of more or less generic abstraction mechanisms do exist (see, e.g., Frantz, 1995; Fishwick and Lee, 1996; Timpf, 1999; Holte and Choueiry, 2003). However, due to their close relationship to the two general hierarchy concepts (cf. Chapter 2.1.2), *generalization* and *aggregation* are usually considered being the most important methods of abstraction.

"Generalization refers to an abstraction in which a set of similar objects is regarded as a generic object." (Smith and Smith, 1977, p. 106). Generalization and the converse process of *specialization* are those processes that produce a classification hierarchy, i.e., elements are hierarchically ordered by "is-a" relations. On the contrary, *"aggregation is an abstraction which turns a relationship between objects into an aggregate object."* (Smith and Smith, 1977, p. 105). The utilization of this kind of abstraction yields a composition hierarchy with "part-of" relations among different levels. Hence, the converse process of aggregation is called *decomposition*. Both major kinds of abstraction, i.e., aggregation and generalization, form a hierarchy of levels. Therefore, organizational and abstraction levels typically coexist within the same hierarchy and that is why abstraction plays an important role in hierarchical multilevel modeling.

A need for multiple abstraction levels often emerges automatically due to observations made at different levels. To appropriately reflect such different observations, parts of the model may have to be specified differently with respect to the grade of abstraction, i.e., the level of detail. For example, in ecology and sociology, the study of the behavior of individuals (micro level) is increasingly important (Hogeweg and Hesper, 1990; Troitzsch et al., 1996). However, implications for the whole population (macro level) might be of interest as well, which are determined by collected behavior of many individual entities. Also, by conducting empirical studies, statistically valid observations can oftentimes be obtained at population level only. Hence, certain dynamic processes might be better modeled at the macro level of abstraction. In this case, a combination of micro and macro level dynamics might be desired, where both levels influence each other via upward and downward causation. However, a micro-macro model does not only include different levels in terms of structural relations within a composition hierarchy, but they include also different abstraction levels, as populations denote an aggregation of individuals and thus a more abstract representation. Therefore, modeling at multiple abstraction levels is widely applied in social sciences and ecology (Schillo et al., 2001; Tilly, 1998; Hogeweg, 2007), and field-specific solutions have been developed to facilitate this endeavor, e.g., the MIMOSE modeling language (Möhring, 1996) and EMSY (Uhrmacher, 1995).

Besides such *structural abstraction*, another commonly applied approach in complex systems modeling is *behavioral abstraction* (Fishwick and Lee, 1996; Lee and Fishwick, 1996), where certain dynamic processes are replaced by an approximative description. *"Behavioral abstraction is one where a system is abstracted by its behavior–that is, we replace a system component [here, component refers to a set of dynamic processes] [...] with something more generic which produces similar behavior."* (Fishwick and Lee, 1996, p. 257). Thereby, in contrast to structural abstraction, no entirely different kinds of entities and additional hierarchical levels are introduced (see Figure 2.4). Behavioral abstraction may, for example, eliminate certain intermediate steps of a dynamic process. Also, approximating the kinetic laws of certain dynamic processes and steady-state assumptions fall into the approach of behavioral abstraction.

The motivation for integrating multiple abstraction levels may be based on different aspects. Besides on the availability of data, the chosen level of abstraction depends also largely on the objective of the model: *"different levels of abstraction can provide for a more natural way of modeling and help to focus on different degrees of detail when using a model"* (Daum and Sargent, 1999, p. 1476). The central question is: how much detail is needed and what are the basic entities of interest? Depending on the answer, a need for different abstraction levels might then automatically come up.

In addition to the model's objective and the availability of data, there is another common reason for introducing multiple abstraction levels, namely to simply reduce model complexity in terms of computing performance (Holte and Choueiry, 2003). Regarding the above sociological example, a model might

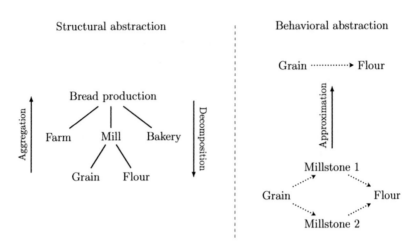

Figure 2.4: Difference between structural and behavioral abstraction. A structural abstraction process forms a hierarchy consisting of different kinds of entities at each level. Behavioral abstraction is an approximation of behavior among certain model entities.

become easily too complex to be simulated in reasonable time, when modeling large populations at the detailed micro level only. Thus, to enhance simulation speed, one might want to abstract from certain processes. It might be sufficient to describe only few processes at the detailed micro level, while the majority is appropriately modeled at the coarse-grained macro level, which is typically much more efficiently to simulate. Choosing an appropriate level of abstraction may help to reduce model complexity and thereby enhancing the performance of simulations (Chwif et al., 2000), but without the ability to combine different abstraction levels, finding an appropriate level of detail can be a hard task and may easily lead to oversimplification and thus to an invalid model. By combining different abstraction levels within a single model, the risk for oversimplification can be reduced as processes can be described at different levels of detail.

2.3 Why Systems Biology Calls for Multilevel Modeling

The major goal of systems biology is to understand functional characteristics and design principles of biological systems (Kitano, 2002b; Kirschner, 2005; Ferrell, 2009). In the past decades – from one of the first computational simulation studies in the mid-twentieth century by Hodgkin and Huxley (1952) until today – systems biology research has lead to the development of an uncountable number of models. Most of them are employing a flat modeling approach and thus questions regarding the requirement for multilevel techniques in systems biology might arise. In the following, a discussion on the benefits of multilevel modeling and on reasons for an increasing need for according approaches tries to answer those questions.

2.3.1 Organization of Biological Systems

Most systems biology research is focusing on the intracellular aspects of metabolic and signaling networks, where certain system properties are emerging from complex molecular interactions. Emergent properties one might be interested in comprise steady state concentrations of involved proteins or the emergence of oscillatory gene expression patterns, for instance. Although systems biology is – due to the analysis of complex *dynamics* among *various* components – typically considered being different from classical reductionistic research, many existing models benefit from and are to a large extent in the line with a reductionistic approach. Their objective is the understanding of emergent systems behavior, e.g., the differentiation of a stem cell, while the focus lies on properties of small interacting constituents, e.g., proteins. Hence, *"systems biology stands to gain a lot from reductionism, and in this sense systems biology is anything but the antithesis of reductionism"* (Ferrell, 2009).

On the other hand, however, several examples previously given in this chapter indicate that biological systems are hierarchically organized and that higher level behavior may influence dynamics at lower levels. Indeed, *"the principle of hierarchic order in living nature reveals itself as a demonstrable descriptive fact"* (Weiss, 1969, p. 4) and both directions of causalities, i.e., upward

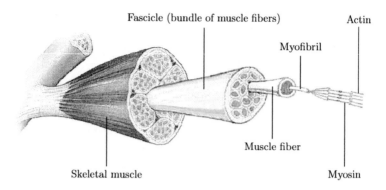

Fascicle (bundle of muscle fibers)   Actin

Myofibril

Muscle fiber

Skeletal muscle   Myosin

Figure 2.5: Hierarchical composition of a skeletal muscle.[1]

causation as well as downward causation, are intrinsic properties of biological systems (Campbell, 1974; Noble, 2006).

For example, highly-developed living systems like vertebrates and similarly advanced animals are hierarchically organized into organs, tissues, and cells: each organ is made of functional tissues, which in turn are composed of interacting cells, i.e., the basic biological building blocks. As this is just a coarse and rather general organization scheme, many further levels can be distinguished when looking at concrete systems in more detail. For instance, muscles are composed of several interconnected bundles (fascicles) of muscle fibers. Each fiber contains several myofibrils and each of them is composed of myofilaments of actin and myosin in turn (Figure 2.5). Anyway, the hierarchical composition of biological systems seems to be an obvious actuality, but how about different levels influencing each other as stated above?

Let us consider the muscle example in more detail. The contraction of a muscle is the result from hundreds of millions or even billions of individual movements of actin against myosin. The physical forces of these low-level protein micro-movements are propagating along the hierarchy, i.e., many very small movements at the nanometer scale are summing-up through connected myofibrils, muscle fibers, and multiple bundles of fibers, so that the entire muscle finally makes a rather large contraction at the macroscopic scale of

---

[1]Modified illustration by the U.S. National Cancer Institute at the National Institutes of Health (http://training.seer.cancer.gov/anatomy/muscular/structure.html)

centimeters. In this case, causation between levels is clearly directed upward. But what has initially caused this process?

At the level of myofilaments, i.e., actin and myosin proteins, a crucial event for causing micro-movements is the release of calcium ions from the sarcoplasmic reticulum. However, the calcium release results from propagating action potentials along the axons of motoneurons. Such neuronal signals originate at once higher levels of the central nervous system, oftentimes due to certain processes in the brain that have been initiated by environmental stimuli, e.g., certain cognitive processes caused by the visual perception of a dangerous situation. Hence, the initial cause for low-level calcium release within muscle fibers can be found at higher levels of organization and thus the causal chain is directed downwards the hierarchy.

However, the entire dynamic process – i.e., muscle-driven actions after perceiving a danger, like defending against an attacker or taking refuge – involves both, downward and upward causation. At first, high-level processes of the central nervous system trigger low-level calcium release via downward causation. The resulting low-level movements of proteins then initiate a cascade of upward causations which finally leads to the high-level contraction of skeletal muscles. Therefore, considering bidirectional relations between multiple levels may be necessary for understanding the full story, i.e., flat single-level models seem to be unsuitable to get deeper insights into general principles of life or the functioning of certain biological systems, as they do not take context appropriately into account. *"The biological systems are, in general, multi-level [...] systems and the attempt to represent them as a single-level (even if multi-variable) system [...] might lead to a model which is valid only over particularly narrow sets of conditions."* (Mesarović, 1968, p. 69f).

Although there is no doubt about the hierarchical nature of biological systems, at first view, modeling systems like the above example, i.e., truly large hierarchies with four and more organizational levels that are spanning several magnitudes of scale, may seem to be too complex and thus impossible to be integrated within a single model. Also, one might argue that multilevel modeling has only limited relevance for systems biology in general, as most research in this field mainly focuses on molecular level processes, which are still barely understood and sufficiently complex on their own. Introducing further levels would make the whole story even more complicated and would be therefore typically not helpful nor desired, if the subject of interest is bounded to molecular processes. However, both skepticisms can be easily refuted.

The first argument against multilevel modeling can be confronted by already existing models, which demonstrate the successful integration of numerous hierarchical levels with highly varying scales in space and time. For instance, probably the most prominent example and among the first large-scale multilevel modeling projects is a series of models describing a virtual heart from genome level up to the entire organ (Noble, 2002; Hunter et al., 2003, 2008; Lee et al., 2009). Other advanced examples comprising multiple integrated levels of organization are dedicated to modeling tumor growth (Zhang et al., 2007; Hirsch et al., 2010; Cristini and Lowengrub, 2010), epithelial tissue renewal (van Leeuwen et al., 2009; Adra et al., 2010), or complex plant processes (Pradal et al., 2008; Band and King, 2012). So, multilevel modeling has proven to be valuable and applicable for diverse biological system. Thereby, the intention for introducing multiple levels typically originates in a combination of dynamic processes operating at highly varying spatial and temporal scales. However, this does not mean that multilevel approaches are useless when modeling systems at the molecular level only.

Hierarchical organization can be found at any scale and biological systems show a rather fine-grained variable structuring at the level of cells and molecules. For example, besides the cytoplasm, especially eukaryotic cells contain various distinct organelles and nested membrane-bound compartments like the nucleus, the endoplasmic reticulum, the Golgi apparatus, lysosomes, and numerous other membrane-bound vesicles (Figure 2.6). Thereby, for proteins and other molecules, a cell is partitioned into multiple hierarchically nested reaction compartments, each of them denoting possibly different conditions. Also one level down the hierarchy, i.e., at the level of molecules, multiple levels might help to structure the model. Many proteins, for instance, are composed of different functional subunits that are interacting with each other and multiple proteins may collectively form large protein complexes that show entirely new behavior compared to the solitary molecules.

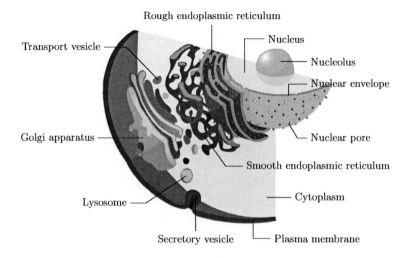

Rough endoplasmic reticulum

Transport vesicle

Nucleus

Nucleolus

Nuclear envelope

Nuclear pore

Golgi apparatus

Smooth endoplasmic reticulum

Lysosome

Cytoplasm

Secretory vesicle

Plasma membrane

Figure 2.6: Hierarchical multi-compartmental structure of a living cell. Cut through the endomembrane system.[2]

Therefore, although systems biology research is still mostly dedicated to investigating the dynamics at a single level of molecules or cells rather than to the big whole, i.e., the complex interplay between a multitude of different levels from molecules up to entire organs and organisms, multilevel techniques might already be useful for rather small models focusing on a relatively narrow band around the molecular scale.

We can conclude: Various hierarchical levels can be distinguished within biological systems. Thereby, feedback loops and other interactions do not only exist between components at the same level, but also across different levels. Thus, to appropriately reflect the complex hierarchical nature of life, models of such systems may need to include multiple levels as well. *"There is [...] no alternative to copying nature and computing these interactions to determine the logic of healthy and diseased states."* (Noble, 2002, p. 1678). Accordingly, multilevelness has been identified to be an important principle of

---

[2]Modified illustration by Mariana Ruiz Villarreal (LadyofHats), public domain, via Wikimedia Commons (http://commons.wikimedia.org/wiki/File:Endomembrane_system_diagram_en.svg).

systems biology, compactly summarized by Denis Noble (2008, p. 17) with a few strong words: *"Biological functionality is multilevel."* Hence, the models should be multilevel too.

2.3.2 Availability of Heterogenous Data

The employment of multilevel modeling methods might be motivated by different aspects. Besides the general and most important incentive, i.e., the hierarchical perception of biology and the need to incorporate dynamics at multiple levels into models in order to understand certain system behavior, another motivation might originate from the availability of experimental data.

Depending on the applied experimental technique, different kinds of observations can be made, i.e., data from different hierarchical levels can be obtained: *"Historically, some molecular biologists have preferred to work at the level of atomic-resolution structure and others at the level of whole-animal physiology. Both are equally valid and there is no reason to believe that this divergence of interest will disappear."* (Sorger, 2005, p. 10). Driven by the ideas of general systems theory, soon the perception evolved that each level has its own being and thus can yield a useful model of the according dynamic processes: *"Most biological phenomena can be examined at many different magnifications. At each magnification interesting and even useful observations can be made and often prediction of future behavior or even correction of malfunctions can also be made."* (Bradley, 1968, p. 39).

Intense progress in the development of new experimental techniques during the past decades strongly promoted the advance in systems biology (Kitano, 2002a,b). Today, experimental biologists have a large set of wet-lab and imaging techniques at hand to collect data from diverse levels of organization. For example, at the level of single molecules, methods like X-ray crystallography and nuclear magnetic resonance (NMR) spectroscopy can be used to determine the three-dimensional structure of proteins. Important techniques for the identification of protein-protein interactions and the according binding rates are immunoprecipitation (IP), yeast two-hybrid screening (Y2H), and surface plasmon resonance (SPR). Numerous methodologies of functional genomics and proteomics, like microarray analysis, immunoblotting, 2-D gel electrophoresis,

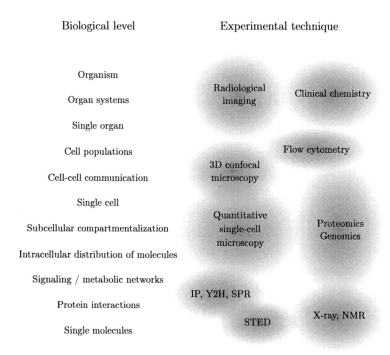

Figure 2.7: Relationships between biological levels and wet-lab techniques.[3]

and mass spectrometry, help to determine the structure and dynamics of intracellular reaction networks. Flow cytometry is an important method for the high-throughput quantification of whole cell populations, and at even higher levels of organization clinical chemistry and certain imaging techniques come into play, which comprise diverse radiological methods, magnetic resonance imaging (MRI), and microscopy (Kherlopian et al., 2008). The latter can be also used to observe spatial structures and dynamics at cell and subcellular levels, down to the nanometer scale with stimulated emission depletion (STED) microscopy. Figure 2.7 illustrates the diversity of experimental approaches and their usage for obtaining data from certain hierarchical levels of biological organization. For a broad and more comprehensive overview of important experimental techniques see also Klipp et al. (2005, Chapt. 4).

---

[3]Modified scheme from Meier-Schellersheim et al. (2009).

The more data are available, the more complex the process of modeling may become, as ignoring certain experimental data may lead to an invalid model. Thus, there is a strong relation between data and model: *"The model must be 'appropriate' to the available data. It must contain variables that connect to all the available observations on genes, proteins, metabolites, etc. If the model is too simple, it will not be able to account for the available data. If it is too complex, there will be insufficient experimental observations to constrain the model."* (Sauro et al., 2006, p. 1725).

Multilevel modeling may help at that point, as it involves the integration of heterogenous data from different levels of organization, potentially comprising processes of different scales and abstraction levels. A frequently found approach is, for example, the combination of phenomenological observations at cell level and detailed mechanistic processes of biochemical reactions (Meier-Schellersheim et al., 2009). However, *"almost any biological experiment will produce data that can potentially be incorporated into a multiscale model"* (Meier-Schellersheim et al., 2009, p. 11), or more generally, into a multilevel model. Therefore, the increasing availability of high-quality data due to the steadily growing and improving repertoire of wet-lab techniques, especially for high-throughput experiments and for observing spatial structures and temporal processes in high resolution, is clearly one of the main driving forces for an increasing interest in multilevel modeling.

2.3.3 Complexity and Accessibility of Models

Another motivation for multilevel modeling in systems biology is to enhance the accessibility of models by reducing their complexity.

Model complexity has many facets, to those belong the complexity of the underlying data, the level of detail, and the accuracy of predicting certain system behavior (Wallace, 1987; Edmonds, 1999, 2000; Chwif et al., 2000). Defining exact measures of complexity is difficult as most aspects of model complexity are part of the so called "psychological complexity" (Curtis, 1980; Wallace, 1987), i.e., complexity depends on the human being who is working with a particular model. Thus, the complexity of a model is strongly related to the language with which the model is described.

Definition *"Complexity is that property of a model which makes it difficult to formulate its overall behaviour in a given language, even when given reasonably complete information about its atomic components and their inter-relations."* (Edmonds, 1999, p. 72).

The language of representation should hence be carefully chosen both with respect to its appropriateness for describing the modeled system but also its understandability for people who work with the model.

As systems biology is a highly interdisciplinary field of research lying at the edge between experimental wet-lab biology and computer science (Kitano, 2002b,a), the design of modeling languages here needs special attention. The technical qualification for describing a certain system may only be half the battle. A carefully designed modeling language may help experimentalists to understand what has been formally described by modelers and thus may improve communication between the different disciplines. Furthermore, as biologists are typically not familiar with the technical details of computational modeling, an accessible formal language is crucial for a broader usage of modeling and simulation methods by biologists themselves (Keane, 2003; Faeder, 2011). Hlavacek et al. (2006, p. 15) point out that *"there is a clear need for making the modeling process more accessible to nonspecialists"*.

The ease of use of a particular modeling language depends on many different *cognitive dimensions* (Green and Petre, 1996). "Closeness of mapping" is one of them and refers to the straightforwardness of transforming an informal mental or conceptual model into a formal representation. Multilevel modeling helps to structure the knowledge about hierarchical systems by providing explicit means for hierarchies and dynamic relations between different levels (Luna, 1993; Chwif et al., 2000). Biology is hierarchically organized (see Chapter 2.3.1) and biologist are used to think this way. Therefore, an easily understandable multilevel language may reduce the psychological complexity of modeling in systems biology and thereby enhance its accessibility.

As has been already said, due to the psychological aspects, the overall complexity of a given model is hard to determine: *"psychological complexity can neither be conclusively defined nor precisely measured. What is measurable are the properties of a model representation that are believed to reflect psy-*

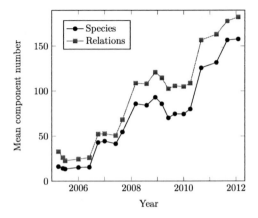

Figure 2.8: Increasing average size of models in the BioModels Database.[4] Relations include different kinds of dynamic relationships like reactions, rate rules, events, and assignment rules.

chological complexity" (Wallace, 1987), like the size of a model description. Characteristics like lines of code or, in the case of biochemical models, the number of species and reactions, are considered to affect the manageability and understandability of a model (Sauro et al., 2006; Hlavacek et al., 2006).

In recent years, increasingly large biological models can be observed. For example, the BioModels Database (Le Novère et al., 2006; Li et al., 2010) – a popular model repository for systems biology – shows an increase of the average number of species and reactions per model by roughly a factor of ten within the last six years (Figure 2.8). The largest models consist of thousands of species and reactions, e.g., a reconstruction of the human hepatocyte metabolic network[5] (Gille et al., 2010), and already much smaller model descriptions challenge their manageability. In Sauro et al. (2006), for instance, Shaffer and Tyson made a statement on a model comprising 30 differential equations (each representing a molecular species) and 140 rate constants: *"This level of complexity stretches the ability of experienced and dedicated modelers to build, analyze, simulate, verify, and test their models by hand."* (Sauro et al., 2006,

---

[4]21st release from February 2012. Source of statistical data: http://www.ebi.ac.uk/biomodels-main/static-pages.do?page=release_20120208.

[5]BioModels entry: http://www.ebi.ac.uk/biomodels-main/MODEL1009150000.

p. 1725). The question is, how does hierarchical multilevel modeling help to reduce complexity in terms of size of model descriptions?

One way to reduce the complexity of models is to employ a *divide and conquer* approach, i.e., modularization and composition: *"ultimately models will be too complex to understand without some form of structuring into units. Model composition, where models are decomposed into structural parts, will be required. Such parts must be understood in terms of their interfaces to other parts."* (Sauro et al., 2006, p. 1726). Although the entire composed model is typically larger than an equivalent monolithic description, individual modules or components may be much more compact and thus more easy to understand (Daum and Sargent, 1999). However, such component-based modeling is only loosely coupled to multilevel modeling as has been introduced in Chapter 2.2, i.e., model composition may be facilitated by hierarchically nested model structures but component-based modeling does not necessarily denote a prerequisite for multilevel modeling nor imply a support for upward and downward causation (Maus et al., 2008). Furthermore, model composition needs to face its own challenging problems, e.g., regarding model validity and viable interface descriptions (Röhl, 2008; Himmelspach et al., 2010).

Already discussed earlier (cf. Chapter 2.2.4), choosing the level of abstraction differently for different subsystems may lead to a less complex model, as parts of it omit a detailed description (see also Uhrmacher et al., 2005). The level of detail determines the atomic entities and has therefore a strong impact on the complexity of a model (Edmonds, 1999). The less abstract parts of a model are, the more detailed entities and according dynamic processes need to be specified and thus typically result in larger (i.e., more complex) formal model descriptions.

Another approach for keeping model descriptions small and manageable is to eliminate redundancy. In particular at the cellular and subcellular scale of biological systems, different hierarchical levels may contain similar parts, as has been shown in Chapters 2.1.2 and 2.3.1. For example, although different compartments within a cell like the cytosol, the nucleus, and endosomes, are hierarchically nested and thus reside at different levels of organization, they all enclose partly identical protein molecules taking part in the same kinds of reactions. According multilevel models therefore need to describe different hierarchical levels at the same level of detail and consisting of similar entities. Many of the very large models in the BioModels Database are models of such kind, i.e., they consist of various reaction compartments with partly similar species and reactions and thus a lot of redundant information. Also modeling multicellular systems at the tissue level tends to result in large redundancy, as all cells may share the same potential cellular states, constituents, and dynamic processes.

To reduce redundancy within model descriptions, different strategies can be employed. One frequently applied approach is an *object-centered* model design (also known as the concept of *process interaction*; see, e.g., Hooper (1986) and Balci (1988)). Focusing on objects is widely used in the context of hierarchical systems modeling, e.g., in DEVS (Zeigler, 1987). By constituting this world view, *"a modeler describes the life cycle of an object which moves through and interacts with the processes of the system under study"* (Balci, 1988, p. 293). Concrete objects are characterized by different attributes. By defining general classes of model components and instantiating multiple similar objects from the same class, the size of model descriptions may be effectively reduced compared to languages that lack such capability (Hooper, 1986). *"The use of objects therefore improves the reliability and maintainability of system code."* (Derrick et al., 1989, p. 713).

In systems biology, by contrast, the alternative strategy of *rule-based* modeling attracts more and more attention recently (Hlavacek et al., 2006; Blinov and Moraru, 2012). Here the focus lies on reactions rather than objects, which particularly permits to describe biochemical systems in a straightforward manner. A rule-based model design follows the *activity scanning* world view of conceptual modeling and simulation strategies (Hayes-Roth, 1985; Hooper, 1986; Balci, 1988), which *"produces a simulation program composed of independent modules waiting to be executed"* (Balci, 1988, p. 291). A model based on the activity scanning paradigm therefore *"consists of components which engage in activities, subject to specified conditions"* (Hooper, 1986, p. 154). By employing a schematic rule-based approach, where those conditions are described in terms of conditional patterns, redundancy may be effectively reduced, as activities may be instantiated within different contexts. Furthermore, the activity scanning approach is typically considered to be user-friendly, as models tend to be modular, modifiable, and easy to understand (Balci, 1988).

To summarize, models of biological systems become increasingly complex, which may seriously hamper their implementation and maintainability. Providing accessible modeling languages and making models more understandable may help to significantly diminish some of the problems induced by increasing complexity. One important point for achieving this goal is to reduce psychological complexity according to diverse cognitive dimensions, e.g., by providing means for structuring models similar to our perception of hierarchical relationships in nature. Another point is to reduce complexity in terms of size of model representations, in particular by combining different levels of abstraction and by applying different methodological paradigms, i.e., object-centered and rule-based approaches, to eliminate redundancy within model descriptions. Several different approaches and their capability to reduce model complexity will be discussed in detail in Chapters 4, 5, 6, and 7.

Chapter 3

Recurring Example Model

In this chapter, a rather simplistic toy system is presented serving as a running example throughout the thesis to illustrate strengths and limitations of different modeling approaches in describing certain phenomena. Before discussing different concrete formal descriptions in subsequent chapters, here an informal conceptual model description will be given at first.

The example comprises dynamic processes at the levels of molecules, cells, and cell populations and is intended to incorporate different aspects that are considered to be representative for multilevel models at these levels of organization. Just like the novel approach proposed later in this thesis (Chapter 6), the example does not give attention to levels beyond, since most systems biology research is dedicated to biochemical and cellular processes and at least for them it needs to be investigated, which methods facilitate the endeavor of modeling at multiple integrated levels.

The chapter is structured as follows: In Chapter 3.1, biochemical dynamics of the example model will be explained along with a brief introduction to the modeled system. Dynamic processes at the cellular level are subject of Chapter 3.2 and in Chapter 3.3 the example model will be extended by dynamic relationships between different levels, i.e., between certain subcellular biochemical processes and processes at the cellular level. Finally, a short summary of the model's multilevel aspects will be given.

3.1 Inter- and Intracellular Processes

3.1.1 Cell-Cell Communication via Notch Signaling

Notch signaling describes a family of highly conserved signaling pathways taking part in local cell communication. Due to its role for regulating gene expression in numerous different kinds of developmental differentiation processes and adult cell fate decisions, Notch signaling denotes one of the most important and widespread mechanisms of direct cell-to-cell interaction (Artavanis-Tsakonas et al., 1999; Lai, 2004). At higher organizational levels like tissues, Notch signaling causes complex pattern formation (Meinhardt and Gierer, 2000; Pourquié, 2003) and regulates the proliferation of cells, e.g., in regenerative processes like wound healing (Chigurupati et al., 2007). Thus, Notch signaling qualifies well for an exemplary multilevel model, as the pathway is involved in complex mechanisms that are regulated by interacting processes at different levels of organization. However, before describing how different levels are interconnected and influence each other, at first an introduction of the essential molecular components and processes will be given.

Central components of each Notch signaling pathway are a Notch-type receptor and a Delta-type ligand. Both molecules are transmembrane proteins with an extracellular domain consisting of epidermal growth factor (EGF) repeats (Lai, 2004). The initial process and one of the key events of Notch signaling is an activation of the Notch receptor by binding a ligand molecule via specific EGF repeats. Due to the direct interaction and membrane-anchoring of both the receptor as well as ligand proteins, signal transduction is mediated between adjacent cells only, i.e., Notch signaling is a variant of juxtacrine intercellular communication (see Figure 3.1). Thereby, cells that are expressing Delta act as signal senders, while Notch expressing cells receive the signal (Schweisguth, 2004). Simplified one could say, through its Notch receptors, a cell may sense the amount of Delta protein of a neighboring cell.

3.1.2 Compartmentalized Subcellular Dynamics

Once Notch has been activated by binding Delta ligand, a cascade of intracellular processes transmits the signal from the extracellular space into the nucleus,

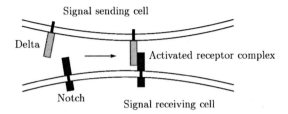

Figure 3.1: Receptor activation in the Notch signaling pathway. A membrane-anchored ligand (Delta) binds to the extracellular domain of an adjacent cell's transmembrane receptor Notch.

where it regulates the transcription of certain target genes. Thereby, different involved proteins and processes are spatially separated by three distinct nested compartments: the plasma membrane, the cytoplasm, and the nucleus.

At the beginning of the signal-transmitting cascade, different protease enzymes are catalyzing the cleavage of activated Notch receptors within the cellular membrane. Proteolytic cleavage releases the extracellular (Necd) and intracellular (Nicd) domains of Notch from its intermediate transmembrane domain. While Necd is bound to Delta and will be degraded subsequently, the intracellular domain Nicd may translocate via the cytoplasmic compartment to its final destination, i.e., the nucleus.

Inside the nucleus, Nicd binds to a transcription factor of the CSL family and acts as a transcriptional co-activator of certain target genes. In the absence of Nicd, CSL binds a co-repression complex, thereby actively inhibiting the expression of Notch target genes. Displacement of the CSL co-repressor by Nicd therefore switches the expression of certain genes from off to on. In addition to such target gene activation, nuclear Nicd may also regulate the Notch signaling pathway itself by positive and negative feedback mechanisms (Artavanis-Tsakonas et al., 1999; Lai, 2004). During *C. elegans* gonadal development, for example, Notch signaling is regulated such that a signal receiving cell up-regulates receptor expression while the expression of Delta ligand is inhibited by Notch activation (Wilkinson et al., 1994). This kind of feedback mechanism leads to lateral inhibition, i.e., neighboring cells will become disparately differentiated, where one cell is predominantly expressing Notch and

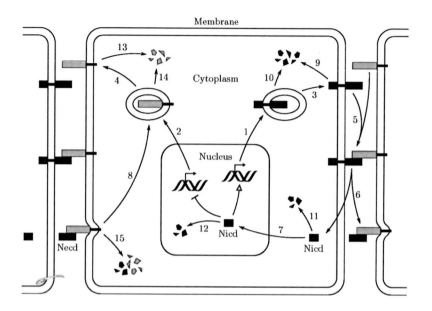

Figure 3.2: Notch signaling model – diagrammatic overview of inter- and intracellular biochemical dynamics.

the other expresses predominantly Delta protein.

Further complex regulatory mechanisms of Notch signaling are controlled by posttranslational modifications, like cleavage and glycosylation of Notch within the Golgi apparatus (Baron et al., 2002). Also, receptors and ligands may be recycled after endocytosis (Schweisguth, 2004; Bray, 2006).

3.1.3   Biochemical Processes

Based on the previously introduced general characteristics of Notch signaling, the example model in this thesis shall include multiple individual cells that may communicate with each other. However, each cell has its own intracellular dynamics in the form of a feedback mechanism, which is schematically illustrated in Figure 3.2 and described in terms of chemical reaction equations in Table 3.1. Dynamic processes at the cellular level and the actual spatial relationships between distinct cells are subject to Chapter 3.2.

Table 3.1: Inter- and intracellular biochemical reaction equations of the Notch signaling model. Molecular species that belong to an adjacent cell are underlined. An asterisk (∗) denotes protein degradation, and mass action, Hill-type, or Michaelis-Menten kinetics are denoted by (MA), (H), or (MM) respectively.

No.	Reaction equation	Kinetic law
1)	GeneN $\longrightarrow$ GeneN + Notch$_c$	(H)
2)	GeneD $\longrightarrow$ GeneD + Delta$_c$	(H)
3)	Notch$_c$ $\longrightarrow$ Notch$_m$	(MA)
4)	Delta$_c$ $\longrightarrow$ Delta$_m$	(MA)
5)	Notch$_m$ + $\underline{\text{Delta}_m}$ $\longrightarrow$ Notch-$\underline{\text{Delta}}$	(MA)
6)	Notch-$\underline{\text{Delta}}$ $\longrightarrow$ Nicd$_c$ + $\underline{\text{Delta}}$-Necd	(MM)
7)	Nicd$_c$ $\longrightarrow$ Nicd$_n$	(MA)
8)	Delta-$\underline{\text{Necd}}$ $\longrightarrow$ Delta$_c$	(MA)
9)	Notch$_m$ $\longrightarrow$ ∗	(MA)
10)	Notch$_c$ $\longrightarrow$ ∗	(MA)
11)	Nicd$_c$ $\longrightarrow$ ∗	(MA)
12)	Nicd$_n$ $\longrightarrow$ ∗	(MA)
13)	Delta$_m$ $\longrightarrow$ ∗	(MA)
14)	Delta$_c$ $\longrightarrow$ ∗	(MA)
15)	Delta-$\underline{\text{Necd}}$ $\longrightarrow$ ∗	(MA)

At the biochemical (or molecular) level, the basic model entities are Notch and Delta. In each cell, both proteins are genetically encoded and thus may be constitutively expressed. Since posttranslational modifications are neglected in this example model, newly synthesized Delta and Notch molecules first reside in the cytoplasm (reaction equations 1 and 2), wrapped by small vesicles preventing from protein interactions within this compartment and leading to subsequent fusion with the plasma membrane, whereby receptor and ligand molecules will become part of the membrane compartment (3–4). To indicate different cellular locations of a protein, in the biochemical reaction equations subscripts m, c, and n are used for denoting the intracellular compartments membrane, cytoplasm, and nucleus respectively. We assume a spatially homogeneous chemical solution within each compartment.

Intercellular signaling and the proteolytic cleavage of activated Notch (Notch-Delta complex) are described by reactions 5 and 6 respectively. Thereby, two different membrane compartments are involved, i.e., Notch and Delta belong to different but adjacent cells, which is indicated by the underlined molecular species in Table 3.1. Instead of multiple sequential proteolytic reactions, the cleavage of activated Notch is described by a single process here and is thus strongly simplified. Products of the proteolysis are cytoplasmic Nicd and the Delta-Necd complex. While reaction 8 describes endocytosis of Delta-Necd for subsequent ligand recycling, Nicd_c may translocate into the nucleus (7). The remaining reactions 9–15 in Table 3.1 describe degradation of Notch and Delta within different compartments.

With translocation of Nicd into the nucleus, the pathway's feedback loop is almost completed. Nicd_n regulates gene expression of Delta and Notch, such that Notch expression is activated and expression of Delta is inhibited by nuclear Nicd. Thereby, the model shall abstract from detailed gene regulatory processes, like cooperative binding of Nicd and CSL transcription factor to a promoter sequence. The activating and inhibitory effect of nuclear Nicd is therefore described as part of the kinetic rates of according reactions. The kinetic rate of reaction 1 follows a sigmoidal saturation, which is described by the Hill equation

$$\theta = \frac{[\text{Nicd}_n]^h}{K^h + [\text{Nicd}_n]^h} \tag{3.1}$$

where θ denotes the fraction of activated transcription factors, $[\text{Nicd}_n]$ is the amount of nuclear Nicd, K is the amount of Nicd_n that activates half of the transcription factors, and h is the Hill coefficient describing the grade of cooperativity. The rate (or speed) of reaction 1 is then described by $k_1' + k_1\theta$, where k_1' is the basal gene expression rate (in the absence of Nicd activation) and k_1 a characteristic reaction rate constant for activated Notch gene expression. As expression of Delta is inhibited by Nicd_n rather than activated, the rate of reaction 2 is $k_2(1 - \theta)$, i.e., the higher the amount of Nicd_n, the lower is the rate of synthesis of Delta ligand.

Similar to the gene expression processes, reaction 6 is another example of behavioral abstraction. In enzyme theory, a catalytic process like the proteolysis of activated Notch is typically considered to be a two-step process with an intermediate enzyme-substrate complex (Berg et al., 2002, Chapt. 8). Therefore, the entire proteolytic process involves at least three biochemical reactions: the formation and breakage of an unstable enzyme-substrate complex

$$\text{Protease} + \text{Notch-}\underline{\text{Delta}} \; \underset{k_d}{\overset{k_a}{\rightleftharpoons}} \; \text{Protease-Notch-}\underline{\text{Delta}}$$

as well as the final catalytic reaction, i.e., the cleavage of activated Notch

$$\text{Protease-Notch-}\underline{\text{Delta}} \; \xrightarrow{k_{cat}} \; \text{Protease} + \text{Nicd}_c + \underline{\text{Delta}}\text{-Necd}$$

where k_a, k_d, and k_{cat} denote the according reaction rate constants. Please note, like in Table 3.1, the underlined molecular species belong to an adjacent cell. Under certain assumptions, however, the enzymatic process may be approximated and described in a single step, like reaction 6 does. If we assume a constant total amount of enzyme (Protease_{tot}, enzyme conservation law) and also no significant change of the intermediate enzyme-substrate complex due to very fast association and dissociation reactions in comparison to the product formation rate k_{cat} (quasi-steady-state assumption), the overall reaction rate v may then be appropriately described by the Michaelis-Menten equation

$$v = \frac{v_{max}[\text{Notch-}\underline{\text{Delta}}]}{K_m + [\text{Notch-}\underline{\text{Delta}}]} \tag{3.2}$$

where $v_{max} = k_{cat}[\text{Protease}_{tot}]$ and K_m is the Michaelis constant. The Michaelis-Menten equation is a classic example of approximation in biochemistry. For a more comprehensive explanation on enzyme kinetics see, for example, Gutfreund (1995) or Sauro (2011).

The rates of the remaining biochemical reactions (3–5 and 7–15) follow the more detailed kinetic law of mass action. That means, the reaction rate is directly proportional to the amount of reactants. For example, the rate of reaction 3 is $k_3[\text{Notch}_c]$, where k_3 is the rate constant and $[\text{Notch}_c]$ the amount of cytosolic Notch. In the case of protein degradation processes (reactions 9–15), we assume the same rate constant k_{deg} for all reactions.

3.2 Cellular and Cell Population Dynamics

3.2.1 Cell Proliferation

The example model in this thesis shall describe the regulation of cell proliferation via Notch signaling. Therefore, besides the previously described intracellular processes of the signaling pathway, the model consists of multiple individual cells that may not only communicate with each other, but also proliferate.

Proliferation is a combination of cell growth in size and number, i.e., cells become larger and after a while they divide into two daughter cells. Therefore, each individual cell of the model has its own size, which may change over time. However, as a cell consists of different compartments, each of them needs to be described separately. Cells and their enclosed compartments are regarded as three-dimensional objects. Hence, size is described in terms of volume.

For the sake of simplicity, the volume of a nucleus is assumed to be constant, i.e., it does not change over time. A cytoplasmic compartment, by contrast, is assumed to increase exponentially with rate k_g until a volume twice of that at time of birth is reached. Similarly, the size of a membrane compartment is also growing exponentially. However, although assumed to be a three-dimensional compartment like others, the plasma membrane defines the cell's surface and is thus rather thin compared to a cytoplasmic or nuclear compartment. As we assume a spherical cell shape, the volume of the membrane compartment therefore increases more slowly during cell growth compared to the cytoplasm, i.e., its exponential growth rate is assumed to be $k_g/2$.

A cell divides at the point where its cytoplasmic compartment volume has doubled. Thereby, the cellular compartments of both daughter cells are assumed to shrink to the initial volume size of the mother cell, i.e., from one generation to another, cell size homeostasis is ensured.

3.2.2 Spatial Relationships

As Notch signaling involves intercellular communication between adjacent cells, according models often rely on an explicit description of the cellular neighborhood, which is typically realized by modeling discrete spatial positions in either one, two, or three dimensions (see, e.g., Collier et al., 1996; Owen and Sherratt, 1998). To observe the formation of complex patterns due to Notch signaling, a two-dimensional hexagonal lattice has been used by Collier et al. (1996) and Ghosh and Tomlin (2001) for representing cellular tissue structures. However, regularly structured lattices may also comprise other spatial relationships for describing cellular tissues, e.g., a two-dimensional lattice of rectangular shaped cells within a von Neumann neighborhood (Figure 3.3). More detailed multicellular spatial settings are lattice-free representations with continuous coordinates in space (e.g., Meineke et al., 2001; van Leeuwen et al., 2009).

To keep the example model simple, a rather abstract linear (one-dimensional) adjacency of non-migrating cells is assumed. That means, each cell has two neighboring cells at most and once inserted, the position of a cell is fixed. New cells are inserted after cell division. Thereby, one daughter cell remains at the position of the dividing mother cell, while the second daughter is randomly placed at an adjacent position that is not already occupied by another cell.

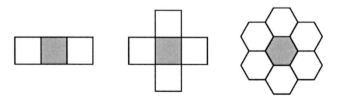

Figure 3.3: Different spatial schemes for modeling lattice-based neighborhoods. From left to right: linear (1D), von Neumann (2D), hexagonal (2D).

3.3 Linking Cellular and Subcellular Processes

The Notch signaling example model includes dynamic processes at different organizational levels, i.e., at the molecular, cellular, and cell population levels. The cellular and cell population levels are interrelated by the process of inserting new cells due to cell division. However, cellular and subcellular processes are totally independent from each other so far. The following section describes two common examples of interlevel causations from subcellular to cellular processes and vice versa.

3.3.1 Upward Causation

In the case of biological multilevel modeling, an exemplary process of upward causation is one where a certain subcellular biochemical condition is influencing certain dynamics at the cellular level. For instance, critical amounts of a certain protein may change the global behavior of a reaction network, e.g., by switching a bistable subsystem on or off respectively. Such emergent behavior may be a marker for crucial cellular states, e.g., different cell fates during stem cell differentiation, which then may influence dynamic processes at other levels of organization. Therefore, in systems biology, multilevel modeling with structural abstraction typically requires to detect certain low-level states for triggering dynamics at higher levels.

It has been shown that Notch receptor activation may cause an arrest of the *cell cycle* and thereby represses cell proliferation (Johnston and Edgar, 1998; Sriuranpong et al., 2001). From one cell division event to the next one, a cell is not just simply growing in size, but it also traverses through different phases of its cell cycle. Thereby, certain checkpoints have to be passed in order to ensure that everything is properly prepared for the next phase. Notch signaling may lead to lateral inhibition, where large amounts of activated Notch cause an arrest in the first (G_1) phase of the cell cycle. The arrested cell may still grow up to a certain volume, but it will not replicate its DNA and therefore no division will take place, i.e., cell proliferation is paused or ultimately stopped.

The example model abstracts from a detailed representation of the cell cycle. We distinguish only between the two growth phases G_1 and G_2, i.e., the DNA

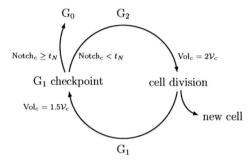

Figure 3.4: Cell cycle dynamics in dependence on cytoplasmic Notch amount. The G_1 checkpoint and cell division are assumed to be at 150% and 200% of the initial cytoplasmic compartment volume respectively.

synthesis phase (S) and mitosis (M) are neglected (Figure 3.4). Also, cell cycle control is assumed to be rather simplistic. We assume the G_1 checkpoint at a cytoplasmic compartment volume $Vol_c = 1.5\mathcal{V}_c$, where $\mathcal{V}_c$ denotes the initial volume of the cytoplasm. The cell may pass this checkpoint and proceed with phase G_2 if the amount of cytoplasmic Notch is low. Otherwise, i.e., if the amount of $Notch_c$ exceeds a threshold value t_N, the cell switches from the *normal* G_1 phase to an *arrested* state (G_0) at that point, which will prevent cell division when its volume has doubled.

3.3.2 Downward Causation

An omnipresent example of downward causation in biological systems is the impact of the volume of a compartment on certain enclosed molecular processes. Membrane-bound cellular compartments are rather elastic objects that may permanently alter their shape and volume. Cells, for instance, are growing and dividing or may change their volume due to osmotic processes.

As the reaction rate of bimolecular and higher order reactions depends on the probability of a collision between randomly moving reactant particles, a larger environmental volume may lower the speed of biochemical reactions. Similarly, a shrinking compartment volume enhances the probability of enclosed particles to collide. In this case, the speed of bimolecular and higher order reactions increases.

Although compartment volumes are changing in the example model, most of the biochemical processes are not affected by this kind of downward causation, as they are first order reactions depending on a single molecular species only. In this case, particles do not need to collide and the speed of a first order reaction is therefore volume-independent. However, reaction 5 in Table 3.1 describes a volume-dependent bimolecular reaction between Notch receptor and Delta ligand. The rate with which both bind together depends on the collision probability. Thus, the according membrane volumes and their alteration may seriously influence the formation rate of activated Notch complexes.

Besides biochemical reactions, translocation processes of molecules from one compartment into another might also depend on the environmental volume, as the mean distance to move increases with larger volumes. However, we assume protein diffusion being very fast compared to the process of passing a membrane barrier. Hence, random movement is not the limiting step in the translocation processes, which are therefore not affected by changing compartment volumes.

3.4 Summary of Multilevel Aspects

The presented example model comprises different aspects that are dedicated to dynamic processes at multiple levels. To a greater or lesser extent, these aspects can be similarly found in many other multilevel models and are therefore considered to be representative for multilevel modeling in systems biology.

Nested hierarchies: The example model comprises of entities (or objects) that are hierarchically arranged in a nested manner. Thereby, identical or similar kinds of entities reside at different locations that are both vertically as well as horizontally separated from each other. For instance, different variants of Notch may reside within different nested compartments of a cell, i.e., at different organizational levels (vertical separation). In addition, the model consists of multiple cells, by which certain entities are separated from each other despite their belonging to the same nesting level (horizontal separation).

Dynamic structures: The structure of the model is changing over time, i.e., structural relationships between entities – like the hierarchically nested composition or bonds between molecules – are not fixed. For example, certain proteins bind and unbind together, they translocate from one compartment into another, and the process of cell division leads to the instantiation of entirely new cells.

State and behavior at different levels: Any level of the hierarchy has a state and behavior of its own. That means, dynamic processes can be found not only at the level of atomic entities that do not enclose further parts, e.g., proteins, but also at higher levels of organization. For instance, the cytoplasm and membrane compartments have a volume state that is changing over time and the entire cell traverses through different phases of the cell cycle. States at different levels may also require to be represented quite differently, e.g., by discrete integers, real numbers, or Boolean values.

Arbitrary rate laws and constraints: Multilevel models often include different rate laws to describe dynamic behavior at different levels of organization. This may also include conditional constraints, such as the restriction that a cytoplasmic compartment in the example model may grow in size by a factor of two at most. Arbitrary rate laws are also useful to make certain behavioral abstractions, like Michaelis-Menten and Hill kinetics instead of the detailed law of mass action.

Upward and downward causation: To describe interlevel causation, a dynamic process at a defined level needs to take information from another level into account, i.e., processes are influenced by certain side-effects that are based on certain states at lower or higher levels of organization. In the case of upward causation, side-effects typically denote an aggregate of states at a lower level, e.g., the total amount of cytoplasmic Notch constrains the cell cycle phase transition in the example model. Conversely, downward causation typically requires to access contextual high-level information, such as the volume of a compartment a process is taking place in.

Space beyond compartmentalization: Besides the compositional hierarchy of entities, multilevel models also often include more detailed spatial aspects. In the example, the topmost level represents a population of cells linearly arranged in a one-dimensional environment, which is a prerequisite to restrict intercellular communication to adjacent cells.

Chapter 4

Flat Model, Multiple Levels?

The following chapter provides a brief and non-exhaustive overview of flat modeling approaches and figures out how certain aspects of biological multilevelness can be described despite the lack of nested model structures. Concrete formal descriptions of the example model presented in the previous Chapter 3 illustrate the suitability of different non-hierarchical formalisms with respect to modeling biological multilevel systems.

The chapter starts with classical popular approaches applied in systems biology, namely differential equations, Petri nets, and process calculi. Chapter 4.2 forges ahead with attributed modeling languages like colored Petri nets and approaches based on rule schemata. The role of constraining model dynamics in a flexible manner will be also discussed in this part. Finally, a short summary of the chapter will be given.

4.1 Classical Approaches

4.1.1 Differential Equations

Traditionally, most dynamic models in systems biology are published in mathematical formulations of ordinary differential equations (ODEs). Nonlinear ODEs are well suited for formally describing biochemical reactions of metabolic and signaling pathways in which molecular substances occur in abundance (Klipp et al., 2005; Ellner and Guckenheimer, 2006). Thereby, different state variables represent the concentrations of distinct molecular species and differential equations describe their dynamic change over time.

For example, the concentration change of cytoplasmic Notch can be described by the following ordinary differential equation (cf. Chapter 3):

$$\frac{d[Notch]}{dt} = \underbrace{k_1' + k_1\theta}_{\text{gene expression}} \overbrace{-k_3[Notch]}^{\text{translocation to membrane}} \underbrace{-k_{deg}[Notch]}_{\text{degradation}} \tag{4.1}$$

where θ denotes the Hill type gene expression kinetics (cf. Equation 3.1 on page 46) and k_1', k_1, k_3, k_{deg} are reaction rate constants of the respective biochemical processes.

An ODE model implicitly assumes each molecular species being located within the same well-mixed reaction compartment and is therefore a prototypical example of a flat modeling approach. Thus, multiple compartments and hierarchical relationships need to be described in an implicit way by annotating according variables and equations. That means, to discriminate between different localities of the same molecular species, each protein within each compartment requires the definition of an own state variable and an according differential equation. For instance, in our example model, Notch receptor molecules reside within the cytoplasm as well as the membrane. Therefore, two different state variables $[Not_c]$ and $[Not_m]$[1] are required to represent the concentration within different compartments. Similarly, Delta protein (cytoplasm and membrane) and the Notch intracellular domain Nicd (cytoplasm and nucleus) also occur within different localities of a cell. Thereby, subscripts n, c, and m denote the nucleus, cytoplasm, and membrane compartments respectively.

Moreover, as the model shall include multiple interacting cells, each of them requires the definition of an own set of ODEs in which different state variables are dedicated to different cells, e.g., via consecutive numbering. As can be easily seen, the number of differential equations for each molecular species increases linearly with the number of different locations the species may reside in. This does not mean to be a problem for relatively small models comprising of just a few compartments, but it may become prohibitive when modeling

---

[1]For the sake of simplicity and due to spatial constraints, protein names may be freely abbreviated throughout the thesis, e.g., Notch and Delta by 'Not' and 'Del' respectively.

multiple cells or more complex intracellular compartmentalization. However, in the case of modeling simple (spatial) relationships showing regular patterns, like the cellular neighborhood in our example model, the number of ODEs may be reduced by formulating corresponding mathematical expressions. For example, the dynamic change of Notch and Delta concentrations within the membrane compartment of a cell can be described as follows:

$$\frac{d[Not_{m,i}]}{dt} = k_3[Not_{c,i}] - k_{deg}[Not_{m,i}]$$
$$- k_5[Not_{m,i}][Del_{m,i-1}] - k_5[Not_{m,i}][Del_{m,i+1}]$$
(4.2)

$$\frac{d[Del_{m,i}]}{dt} = k_4[Del_{c,i}] - k_{deg}[Del_{m,i}]$$
$$- k_5[Not_{m,i-1}][Del_{m,i}] - k_5[Not_{m,i+1}][Del_{m,i}]$$
(4.3)

where $i \in \mathbb{N}$ denotes a numerical index position. Assuming consecutive indices of adjacent cells, the subterms comprising of expressions $i-1$ and $i+1$ properly describe Delta-Notch binding between neighboring cells. However, numerical simulation requires that the tool supports an automated instantiation of concrete ODEs from such schematic descriptions. Alternatively, non-schematic ODEs need to be generated by the user in advance by using high-level or special-purpose programming languages like MATLAB (Lynch, 2004). Hence, the proposed method is a viable solution for compactly representing such kinds of models – e.g., for publication – rather than for actually working with them.

The set of nonlinear ODEs listed in Table 4.1 describes the intra- and intercellular biochemical dynamics within a one-dimensional cellular adjacency, as has been informally described in Chapter 3. In addition, Equations 4.4 and 4.5 describe exponential volume growth of different compartments. Thereby, variables $Vol_{c,i}$ and $Vol_{m,i}$ denote the cytoplasm and membrane volumes of cell i respectively. The volume $Vol_{n,i}$ of nucleus compartments is assumed to do not change in size, i.e., $dVol_{n,i}/dt = 0$.

Besides conventional descriptions of biochemical reactions, the remaining differential equations also include additional expressions to reflect downward causation by altering compartment volumes. However, unlike what might be understood by reading Section 3.3.2, thereby each differential equation is af-

Table 4.1: Set of ODEs describing biochemical processes of the Notch signaling example including dynamic compartment volumes.

$$dVol_{c,i}/dt = k_g\, Vol_{c,i} \tag{4.4}$$

$$dVol_{m,i}/dt = \frac{k_g}{2}\, Vol_{m,i} \tag{4.5}$$

$$d[Not_{c,i}]/dt = k_1'\frac{Vol_{c,i}}{Vol_{c,i}} + k_1\frac{[Nic_{n,i}]^h}{K^h + [Nic_{n,i}]^h}\frac{Vol_{m,i}}{Vol_{c,i}} - k_3[Not_{c,i}] - k_{deg}[Not_{c,i}] - k_g[Not_{c,i}] \tag{4.6}$$

$$d[Del_{c,i}]/dt = k_2(1 - \frac{[Nic_{n,i}]^h}{K^h + [Nic_{n,i}]^h})\frac{Vol_{m,i}}{Vol_{c,i}} + k_8[NecDel_i]\frac{Vol_{m,i}}{Vol_{c,i}} - k_4[Del_{c,i}] - k_{deg}[Del_{c,i}] - k_g[Del_{c,i}] \tag{4.7}$$

$$d[Not_{m,i}]/dt = k_3[Not_{c,i}]\frac{Vol_{c,i}}{Vol_{m,i}} - k_5[Not_{m,i}][Del_{m,i-1}] - k_5[Not_{m,i}][Del_{m,i+1}] - k_{deg}[Not_{m,i}] - \frac{k_g}{2}[Not_{m,i}] \tag{4.8}$$

$$d[Del_{m,i}]/dt = k_4[Del_{c,i}]\frac{Vol_{c,i}}{Vol_{m,i}} - k_5[Not_{m,i-1}][Del_{m,i}] - k_5[Not_{m,i+1}][Del_{m,i}] - k_{deg}[Del_{m,i}] - \frac{k_g}{2}[Del_{m,i}] \tag{4.9}$$

$$d[Not_iDel_{i-1}]/dt = k_5[Not_{m,i}][Del_{m,i-1}] - \frac{v_{max}[Not_iDel_{i-1}]}{K_m + [Not_iDel_{i-1}]} - \frac{k_g}{2}[Not_iDel_{i-1}] \tag{4.10}$$

$$d[Not_iDel_{i+1}]/dt = k_5[Not_{m,i}][Del_{m,i+1}] - \frac{v_{max}[Not_iDel_{i+1}]}{K_m + [Not_iDel_{i+1}]} - \frac{k_g}{2}[Not_iDel_{i+1}] \tag{4.11}$$

$$d[NecDel_i]/dt = \frac{v_{max}[Not_{i-1}Del_i]}{K_m + [Not_{i-1}Del_i]}\frac{Vol_{m,i-1}}{Vol_{m,i}} + \frac{v_{max}[Not_{i+1}Del_i]}{K_m + [Not_{i+1}Del_i]}\frac{Vol_{m,i+1}}{Vol_{m,i}} - k_8[NecDel_i] - k_{deg}[NecDel_i] - \frac{k_g}{2}[NecDel_i] \tag{4.12}$$

$$d[Nic_{c,i}]/dt = \frac{v_{max}[Not_iDel_{i-1}]}{K_m + [Not_iDel_{i-1}]}\frac{Vol_{m,i}}{Vol_{c,i}} + \frac{v_{max}[Not_iDel_{i+1}]}{K_m + [Not_iDel_{i+1}]}\frac{Vol_{m,i}}{Vol_{c,i}} - k_7[Nic_{c,i}] - k_{deg}[Nic_{c,i}] - k_g[Nic_{c,i}] \tag{4.13}$$

$$d[Nic_{n,i}]/dt = k_7[Nic_{c,i}]\frac{Vol_{c,i}}{Vol_{n,i}} - k_{deg}[Nic_{n,i}] \tag{4.14}$$

fected, including those describing first-order reactions only. ODEs describe the amounts of molecular species in terms of real-valued concentrations rather than discrete particle numbers. As concentration is defined by particle number per volume, the value of according state variables always depends on the surrounding compartment volume and therefore needs to be adapted in the case where this volume is dynamically changing.

In the example model, changing volumes influence the concentration of molecular species differently. First of all, concentrations become diluted by increasing compartment volumes. According to the method proposed by Barberis et al. (2007), such volume-dependent concentration changes can be described by mathematical terms like in the following differential equation in the absence of any further dynamics:

$$\frac{d[X]}{dt} = -\frac{[X]}{Vol}\frac{dVol}{dt} \qquad (4.15)$$

where $[X]$ denotes the concentration of a molecular species X that is enclosed by a compartment with volume Vol, and $dVol/dt$ is the dynamic change of Vol over time. In the case of cytoplasmic molecules of the Notch signaling example we can write

$$\frac{[X_{c,i}]}{Vol_{c,i}}\frac{dVol_{c,i}}{dt} = \frac{[X_{c,i}]}{Vol_{c,i}}k_g\,Vol_{c,i} = k_g[X_{c,i}] \qquad (4.16)$$

where $[X_{c,i}]$ is a placeholder for $[Not_{c,i}]$, $[Del_{c,i}]$, and $[Nic_{c,i}]$. In the same way, a dilution of membrane-residing molecules can be described (see according equations in Table 4.1).

The second type of adjusting molecular concentrations in dependence on compartment volumes refers to translocation processes from one compartment into another. As the volume of the former compartment may be different to that of the destination, translocation process terms therefore need to be adjusted by a volume ratio factor, e.g., $Vol_{n,i}/Vol_{c,i}$ in the case of translocation from the nucleus into the cytoplasm.

Let us take a complete example. Equation 4.1 – which describes the change of cytoplasmic Notch within a static compartment volume – needs to be extended as follows in order to include all volume-dependent processes, i.e., adaptations referring to dilution (due to growing compartment volume) as well as translocation (as part of the gene expression process):

$$\frac{d[Not_{c,i}]}{dt} = \underbrace{k_1' \frac{Vol_{n,i}}{Vol_{c,i}}}_{\text{basal gene expression}} + \overbrace{k_1 \theta \frac{Vol_{n,i}}{Vol_{c,i}}}^{\text{activated gene expression}} -k_3[Not_{c,i}] - k_{deg}[Not_{c,i}] \underbrace{-k_g[Not_{c,i}]}_{\text{dilution}} \quad (4.17)$$

Modeling biological systems by formulating differential equations provides a rather large degree of flexibility in describing the model's dynamics. Besides the diverse volume-dependent adaptations and the schematic description of cellular neighborhoods as shown in Table 4.1, Equations 4.6 and 4.7 as well as 4.10–4.13 also include formulations of alternative biochemical reaction kinetics, i.e., Hill-type and Michaelis-Menten kinetics.

Moreover, mathematical formulations allow for integrating processes at highly diverse levels of abstraction, as the modeler may utilize the whole repertoire of mathematics. For example, partial differential equations (PDEs) can be used to model reactions in dependence on concentration gradients (Neves and Iyengar, 2009; Wittmann et al., 2009), and delay differential equations (DDEs) to describe abstract dynamic processes of gene expression (Lewis, 2003) as well as at the cellular level, like cell differentiation (Lai et al., 2009) or running through the different phases of the cell cycle (Bocharov and Rihan, 2000; Villasana and Radunskaya, 2003). Thereby, linking of differently abstract processes is facilitated due to the inherent global scope of variables, which allows accessing each variable anywhere.

However, the other side of the coin is a combinatorial explosion, if certain model components show multiple distinct states or binding partners and reside at different compartmental locations. Also, as explicit structures are missing, it is difficult to make a distinction between different parts of the model with respect to their organizational or abstraction levels. The lack of structuring elements thus may lead to rather large and cluttered model descriptions of limited accessibility.

Impossible to describe with ODEs are discrete events like the cell division process of the Notch signaling example (see Chapter 3.2). Modeling the abrupt resetting of compartment volumes and adding new cells dynamically would require a hybrid mathematical approach, where parts of the model are described continuously while others are discrete (see, e.g., Anderson, 2005). The same holds true for the upward causation dynamics, which involve the discrete transition of the cell from G_1 phase to either G_2 phase or to an arrested state of the cell cycle (G_0 phase) depending on certain conditions.

4.1.2 Petri Nets

Another classical modeling approach in systems biology is to specify models in the formal graphical notation of Petri nets (also termed place/transition nets or P/T nets). Petri nets are directed bipartite graphs, where *place* nodes (circles) represent states or conditions and *transition* nodes (bars or squares) the conversion from one condition into another. Directed *arcs* connect place nodes with transitions, such that certain places define pre- and postconditions for each transition.

The Petri net formalism has been developed in the field of theoretical computer science and qualifies for modeling concurrent systems in diverse application fields. However, as Carl Adam Petri states himself, his original motivation for inventing the P/T net notation was to describe reaction networks of chemical processes and their competition for molecular resources (Petri and Reisig, 2008). In this case, place nodes represent molecules or molecular compounds and reactions are described by according transitions (see Figure 4.1 for a simple example net of a bimolecular reaction). Due to the straightforward mapping from a set of (bio)chemical reaction equations to Petri nets and vice versa (Marwan et al., 2009), the place/transition notation has become rather popular for representing models in systems biology.

In the original Petri net formalism, places may be marked by a discrete number of *tokens*, which will be consumed from places denoting preconditions

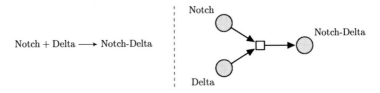

Figure 4.1: Petri net representation of a bimolecular reaction.

(input arcs) and placed at postconditions (output arcs) when a transition fires. However, as an original Petri net has a timeless and non-deterministic execution semantics, its applicability is limited to qualitative modeling and analyses, e.g., analyzing reachability properties or the liveness of the network given a certain marking (Reddy et al., 1993).

To perform quantitative simulation studies, timed Petri nets have been invented, in which the firing of transitions takes a certain amount of time (Wang, 1998). For systems biology, two kinds of timed Petri nets are of particular importance, namely *stochastic* and *continuous* Petri nets (Goss and Peccoud, 1998; Chaouiya, 2007; Gilbert et al., 2007; Heiner et al., 2008a).

The firing of transitions in a stochastic Petri net is chosen randomly according to a stochastic firing rate each transition is assigned with (Goss and Peccoud, 1998). Although stochasticity must not necessarily be restricted to a certain distribution, describing (bio)chemical reaction systems typically implies exponentially distributed transition probabilities, hence the reachability graph of the net is nothing else than a continuous time Markov chain (CTMC). In this case, the stochastic simulation algorithm (SSA) by Daniel T. Gillespie (1977) can be applied to determine which transition fires next and to calculate the time point of this event. Modeling stochastic processes takes the intrinsic noise into account that can be found in many biological systems. For further reading, Wilkinson (2006) provides a comprehensive overview of stochastic modeling and simulation in the context of systems biology.

Similar to stochastic variants, a continuous Petri net has also a timed execution semantics. However, unlike to the original non-deterministic and stochastic Petri nets, places do no longer contain discrete numbers of tokens but represent real-valued variables. Also, transitions of a continuous Petri net do not fire at discrete points in time. Instead, variables are changed simultaneously and constantly, i.e., in a continuous manner. The execution semantics of a continuous Petri net is therefore mathematically defined by ordinary differential equations (Heiner et al., 2008a).

However, in any case and similar to ODE models (cf. previous section), Petri nets require to explicitly define an own state variable, i.e., place node, for each combination of molecular state and location. For example, mature

Notch protein may be located within the cytoplasm and the membrane of a cell. In addition, its intracellular domain Nicd may be located within the cytoplasm and the nucleus. Representing three interacting cells by a Petri net graph therefore requires twelve distinct place nodes for the different variants and locations of the unbound Notch receptor. Similarly, multiple variables are needed to describe the different locations of Delta ligand and complexes comprising Delta and Notch (cf. Figure 4.2). Hence, a Petri net model may contain large amounts of redundancy, as different subnets may describe identical or similar behavior. When modeling multi-compartmental or multi-cellular models, the combinatorial explosion may therefore easily lead to very large graphs consisting of hundreds of nodes (see, e.g., Janowski et al., 2010).

At the cost of burdening nodes with additional information, modeling reaction kinetics beyond standard mass-action biochemistry is supported by specifying arbitrary transition rate functions (Heiner et al., 2008b; Rohr et al., 2010). Thereby, the scope of retrievable information for each transition is typically restricted to the set of pre-places and thus needs to be explicitly specified by connecting defined place nodes.

Howsoever, the expressiveness of elementary (timed) Petri nets seems to be insufficient to describe complex dynamic processes depending on diverse side-conditions, like certain cellular processes or upward and downward causation as described in Chapter 3. For example, a cell may only grow in size as long as the cytoplasmic volume did not reach twice the initial value. Also, at the point of cell division, volumes of both the cytoplasm and membrane compartments shall be reset to their initial size. Another example is the cell cycle transition from G_1 to G_2 phase, which may only fire under the condition that cytoplasmic Notch did not exceed a certain threshold t_N, while reaching the arrested G_0 state requires the opposite way round, i.e., its amount needs to be larger than t_N. To describe such functionalities, Petri net formalisms have been extended by additional syntactical and semantical means, like capacities of places, test arcs (also called read arcs), inhibitor arcs, and reset arcs (see Figure 4.3, see also Marsan et al., 1995; Heiner et al., 2009).

Parts of an extended stochastic Petri net of the Notch signaling example are shown in Figure 4.4. Numerous special arcs (i.e., test, inhibitor, and reset

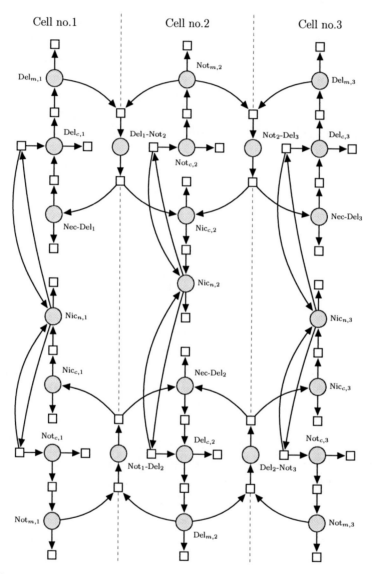

Figure 4.2: Petri net of biochemical Notch signaling processes within a model of three interacting cells. Transition labels and arc multiplicities are omitted.

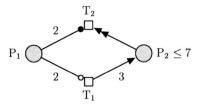

Figure 4.3: Different means of an extended Petri net. Transition T_1 is connected with place P_1 via an *inhibitor arc* that has a *multiplicity* of 2. Therefore, T_1 may only fire if P_1 contains less than two tokens. If T_1 fires, no tokens will be consumed from P_1 and three tokens will be placed at P_2. However, due to the *capacity* of P_2, T_1 may only fire if P_2 does not contain more than seven tokens afterwards. Transition T_2 is connected with place P_1 via a *test arc* and with P_2 via a *reset arc*. T_2 may only fire if P_1 contains at least two tokens. Firing of T_1 does not change the amount of tokens of place P_1, but all tokens of P_2 will be consumed.

arcs) are thereby used to describe different cellular processes and interrelations between cellular and subcellular levels. For example, upward causation is described via guarding the amount of cytoplasmic Notch, i.e., a token number $\geq t_N$ inhibits or enables the firing of two transitions that are denoting the cell cycle's G_1 checkpoint.

A limitation of the stochastic Petri net description is the containment of merely discrete (integer) numbers of tokens. While this denotes an appropriate representation of the amounts of molecular substances as well as different phases of the cell cycle, compartment volumes are typically more properly described by real-valued variables. Therefore, to correctly describe the according processes in a stochastic Petri net, all quantities of the model, like rate constants and molecular amounts, need to be normalized to the smallest volume unit of 1 token. To overcome this limitation, another extension of the formalism has been developed, where different kinds of places and transitions, i.e., discrete and continuous ones, may be part of the same *hybrid* Petri net model (Matsuno et al., 2000, 2003; Proß and Bachmann, 2011).

Another limitation of all of the introduced Petri net variants is the static structure of models. Therefore, none of them allows describing the dynamic creation of new cells after division.

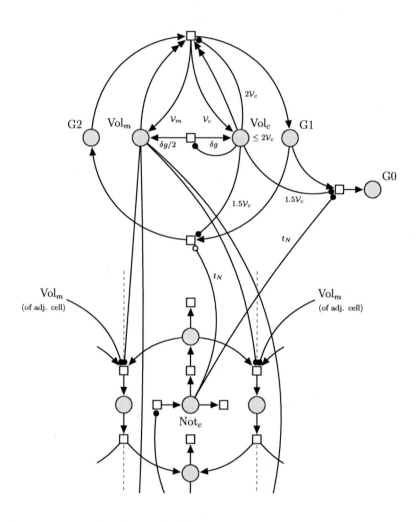

Figure 4.4: Extended stochastic Petri net of upward and downward causation processes within a single cell. Cell indices and transition labels are omitted. Place Vol_c has assigned a capacity of $2\mathcal{V}_c$. Only arc weights larger than 1 are given explicitly.

4.1.3 π-Calculus

The π-calculus by Milner et al. (1992) has been developed for formally describing the changing structure of concurrent communicating systems (see also Milner, 1999). Similar to other process algebras, e.g., CCS (Milner, 1980) or PEPA (Hillston, 1996), its focus lies on the parallel composition and continuation of communicating processes (Baeten, 2005). In the past decade, the π-calculus gained considerable attention in formal computational systems biology since Priami et al. (2001) have suggested its usage and demonstrated the general applicability for modeling biochemical systems.

Just like colliding and reacting molecular particles, the π-calculus denotes an object-centered approach, where certain individual processes interact with each other synchronously and in parallel. Interaction is restricted to pairwise communication via compatible channels, which is a well-fitting analogy to molecular processes relying on complementary structures. Unlike many other process calculi, the π-calculus allows to transmit channel names from one process to another when they are interacting. Such message sending can be used to model molecular modification: *"Chemical interaction and subsequent modification coincide with communication and channel transmission."* (Regev et al., 2001).

Similar to Petri nets, the original π-calculus does not provide a quantitative interpretation. However, the formalism has been extended by a stochastic execution semantics in terms of CTMCs (Priami, 1995). Interactions in the stochastic π-calculus have assigned an exponentially distributed delay, so that the non-deterministic choice of interaction in the original semantics becomes a stochastic race.

The basic rule of action in the π-calculus is the communication of two *processes* by using a name of which both processes share the knowledge of. This name denotes a *channel*. Afterwards, both processes continue with processing their internal "program", finally proceeding with other processes or simply terminating the current ones. A usual abstraction is to describe biochemical species or active molecular motifs (e.g., molecular binding sites) as communicating processes and reactions by pairwise communication over channels (see Priami et al., 2001; Regev et al., 2001; Regev, 2002). For example, the following two processes can describe binding of Notch and Delta proteins:

$$Del \quad ::= \quad bind!.0$$
$$Not \quad ::= \quad bind?.Not\text{-}Del$$

Here, processes *Del* and *Not* represent the Delta ligand and Notch receptor respectively, and *bind* is the name of the channel over which both processes communicate. Thereby, process *Del* initiates the interaction as a sender (*bind!*) and *Not* is the receiving process (*bind?*). The stochastic firing rate of the globally defined communication channel may be specified as $bind@k_5$. Terms following the succession symbol (.) define what will be processed after an interaction occurs. In the case of *Del*, the process will simply terminate (denoted by the idle process **0**), and process *Not* will proceed with another process (*Not-Del*).

Although the paradigm of interacting processes provides a suitable metaphor for describing biochemical reactions, modeling complex biological multilevel systems is hampered due to certain formalism-specific restrictions. To those belongs the fact that actions in the π-calculus are relying on the interaction of exactly two processes. Modeling of unimolecular reactions like protein degradation (Reactions 9–15 in Table 3.1) therefore requires the definition of "timer" or "dummy" *co*-processes (Regev, 2002) acting as communication partners:

$$Not \quad ::= \quad degradation?.0$$
$$Not_{co} \quad ::= \quad degradation!.Not_{co}$$

To avoid this artifact, some variants of the stochastic π-calculus, e.g., the Stochastic Pi-Machine (SPiM) by Phillips and Cardelli (2004), use syntactic sugar that allows for activating processes after exponentially distributed stochastic time delays τ, i.e., internal transitions without communication over a channel. A first-order degradation reaction of Notch can therefore also be described by a single process comprising a stochastic delay at rate k_{deg}:

$$Not \quad ::= \quad \tau@k_{deg}.0$$

However, while the above approach facilitates the description of unimolecular reactions, modeling interactions that depend on more than two entities remains to be impossible in the π-calculus. From a theoretical point of view this might be reasonable, since *"termolecular reactions, involving three molecules in the transition state, are very rarely encountered. Bimolecular reactions have to take place in the very brief time that two molecules collide, before they bounce apart again. The chance that a third molecule will collide at exactly the same time, in a suitable orientation for reaction, is extremely improbable."* (Jackson, 2004, p. 12). However, this applies to elementary chemical reactions only. Since modeling means to abstract from certain details, a combination of multiple elementary reaction steps to one abstract process is common practice in systems biology. Moreover, as has been demonstrated by Kuttler et al. (2010), higher-order reactions *"allow to incorporate global control into models"*, such as defined global conditions or side effects that are also indispensable for multilevel modeling.

Another major limitation of the π-calculus is the restriction to mass-action kinetics. Without the ability to define alternative kinetic laws, modeling biological systems in general and multilevel relationships in particular is hampered due to the inflexibility in specifying a model's dynamics. For example, the catalyzed cleavage of activated Notch (Notch-Delta complex) as well as the gene expression processes of the example model need to be described in more detail than it is desired in order to capture the observed behavior, i.e., multiple intermediate reaction steps instead of abstract Michaelis-Menten or Hill-type kinetics respectively (Kuttler and Niehren (2006) and Regev (2002) present several illustrative examples). Furthermore, the inherent mass-action restriction has a strong impact on the qualification for modeling multilevel dynamics, since incorporating dynamic processes depending on high-level information, like the volume-dependence of certain reaction rates, seems to be very difficult if not impossible to describe.

However, on the other hand the π-calculus provides capabilities to reduce the problem of combinatorial complexity. The formalism allows for restricting the scope of communicating processes, e.g., to model different compartments (Regev et al., 2001). This way, the number of process definitions may be reduced. The communication scope can be restricted by introducing private channels that are not globally known, i.e., of which only a subset of all processes shares the knowledge of. Private channels can be dynamically generated by using the ν operator. The awareness of private channels can be propagated afterwards by transmitting their names between communicating processes and by process parameterization.

For example, the $Init$ process in Table 4.2 starts with creating two private channels id_1 and id_2, which are then used to parameterize two parallel $(\ldots \mid \ldots)$ processes $Cell(pre, this)$. The channels are later used for an interaction between Notch and Delta, such that communication is restricted to adjacent cells (Figure 4.5a illustrates the general principle). That is why both initial cells are parameterized with the same private channels, but in a converse order. Thereby, the first parameter pre denotes an identifier of the predecessor of a cell and the second parameter $this$ an identifier of the cell itself. By generating another private channel suc (denoting its successor), the one-dimensional adjacency of a cell is complete. Due to parameterizing further processes with these private channel names, subsequently instantiated processes of a cell share the same restricted communication scope to control process interaction.

Besides the parameterization of newly instantiated processes, a restricted communication scope can be also achieved by transmitting private channels between communicating processes. This can be used, e.g., to model bindings between individual Notch and Delta processes, as shown in Table 4.2 (see also the illustration in Figure 4.5b). Therefore, parameterized processes Not_m and Del_m may interact via private channels that have once been created by a $Cell$ or $Init$ process and are locally denoted by t, p, and s respectively. Del_m acts as a sender process and transmits a newly generated private channel $cleavage$ to a receiving process Not_m:

$$Not_m(t, \ldots) \quad ::= \quad t?\{cleavage\} \ldots$$
$$Del_m(p, s) \quad ::= \quad (\nu\ cleavage@k_6)\ (p!\{cleavage\} + s!\{cleavage\}) \ldots$$

As each cell in the one-dimensional environment has potentially two neighboring cells, both interactions with the predecessor $(p!)$ and successor $(s!)$ compete with each other in a stochastic race (choice operator $+$). After an interaction has occurred, the Del_m process waits for an interaction via the private $cleavage$ channel, which is triggered by the previously receiving process. Afterwards, the Not_m process will proceed with a Nic_c process and Del_m

Table 4.2: Stochastic π-calculus model of the Notch signaling example including dynamic creation of cells. Please notice, as the focus of interest lies at the parameterization and restricted scope of processes, the catalytic cleavage of Notch as well as gene expression processes are highly simplified and thus do not fully capture the dynamics described in Chapter 3. Likewise, cell division – which is assumed to occur only once for each cell – is also described by a rather simple process.

$Init$	::=	$(\nu \, id_1@k_5, id_2@k_5) \, (Cell(id_1, id_2) \,|\, Cell(id_2, id_1))$
$Cell(pre, this)$	::=	$(\nu \, suc@k_5, dbs@k_a) \, (GenN(this, dbs) \,|\, GenD(pre, suc, dbs) \,|\, Div(this, suc))$
$Div(this, suc)$	::=	$\tau@k_{div}.Cell(this, suc)$
$GenN(t, d)$	::=	$(\tau@k'_1.(GenN(t, d) \,|\, Not_c(t, d))) + (d?\{unbind\}.unbind!.(GenN(t, d) \,|\, Not_c(t, d)))$
$GenD(p, s, d)$	::=	$(\tau@k_2.(GenD(p, s, d) \,|\, Del_c(p, s))) + (d?\{unbind\}.unbind!.GenD(p, s, d))$
$Not_c(t, d)$	::=	$(\tau@k_3.Not_m(t, d)) + (\tau@k_{deg}.0)$
$Del_c(p, s)$	::=	$(\tau@k_4.Del_m(p, s)) + (\tau@k_{deg}.0)$
$Not_m(t, d)$	::=	$(t?\{cleavage\}.cleavage!.Nic_c(d)) + (\tau@k_{deg}.0)$
$Del_m(p, s)$	::=	$(\nu \, cleavage@k_6) \, ((p!\{cleavage\} + s!\{cleavage\}).cleavage?.((\tau@k_{deg}.0) + (\tau@k_8.Del_c(p, s)))) + (\tau@k_{deg}.0)$
$Nic_c(d)$	::=	$(\tau@k_7.Nic_n(d)) + (\tau@k_{deg}.0)$
$Nic_n(d)$	::=	$(\nu \, unbind@k_d) \, (d!\{unbind\}.unbind?.Nic_n(d)) + (\tau@k_{deg}.0)$

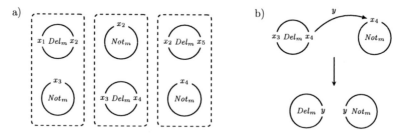

Figure 4.5: Restricted communication scope in the π-calculus. Different communication channels are denoted by different names x, y. (a) Specific parameterization with private channel names x_i guarantees that Delta and Notch processes may communicate with processes of an adjacent cell only. (b) Del_m transmits a newly created name y to the receiving process Not_m. Afterwards, both processes share knowledge of this private channel.

has the choice between degradation ($\tau @ k_{deg}.0$) and recycling to the cytoplasm ($\tau @ k_8.Del_c(p,s)$).

To summarize, the π-calculus provides means for restricting the communication scope of processes. Thereby, the size of model descriptions may be effectively reduced compared to models with global communication channels only, as one generic set of processes may describe dynamics in different contexts. For example, the set of process definitions in Table 4.2 describes the dynamics between arbitrarily large numbers of cells. However, as has been outlined above, certain important aspects of multilevel modeling are difficult if not impossible to describe, e.g., influencing the rate of reactions dynamically via downward causation.

4.2   Attributed Languages

Although the previously introduced approaches follow rather diverse modeling paradigms with different support for describing multilevel phenomena, a common problem they share is combinatorial explosion. Different states of molecules or different locations they reside in typically require the definition of distinct state variables with appropriate names. By contrast, attributed (or colored) languages equip entities with different states and may thereby help to reduce the problem of combinatorial complexity. In the previous section of the π-calculus, the parameterization of processes with different private channel names has already given a first impression of the potentiality for reducing complexity by assigning different parameters (or states) to model entities. However, attributed languages typically go further by allowing to define flexible constraints based on concrete attribute values, as we will see in the following.

4.2.1 Colored Petri Nets

Colored Petri nets (Jensen, 1998; Jensen and Kristensen, 2009) are extensions of classical P/T nets, where token *types* (i.e., attributed tokens) replace conventional tokens. Therefore, each place is assigned a defined data type, e.g., integers $\mathbb{N}$, of which it can hold a discrete number of tokens. Data types may also be Cartesian products of multiple elementary types to represent more than one attribute per token, e.g., $\mathbb{N} \times \mathbb{R}$. Due to different values, tokens in the same place node thus may become distinguishable from each other, rather than just being existent or not. Like for conventional Petri nets, timed variants of colored Petri nets allow for modeling quantitative dynamics.

Conditions assigned to incoming edges of transitions allow constraining their firing based on the color of tokens, i.e., on token attribute values. A transition is enabled if a set of tokens exists that fulfills all conditions of the respective transition. Thereby, conditions can also reference values from other pre-places of the same transition, for instance, to constrain the transition to be enabled only if two tokens from different places have the same values. This way, conditions can be also used to describe binding of Notch and Delta between adjacent cells, as is shown in Figure 4.6. Also more complex spatial relationships can be described by using colored Petri nets, for example, a two-dimensional hexagonal cellular lattice presented by Gao et al. (2011).

Conditions may also rely on more than one attribute per token. For example, place nodes in Figure 4.7 that are representing diffusible protein species are assigned the data type $\mathbb{N} \times \{\mathtt{m}, \mathtt{c}, \mathtt{n}\}$, which describes the index position of the cell and the cellular compartment the protein resides in, i.e., either the membrane ($\mathtt{m}$), cytoplasm ($\mathtt{c}$), or nucleus ($\mathtt{n}$) respectively. For example, conditions $(i, \mathtt{m})$ and $(j = i + 1$ or $j = i - 1, \mathtt{m})$, which are assigned to the incoming

73

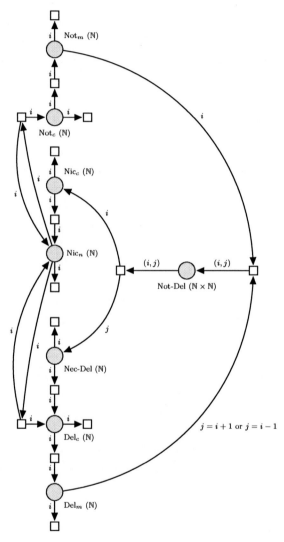

Figure 4.6: Colored Petri net of biochemical Notch signaling processes. Transition labels are omitted.

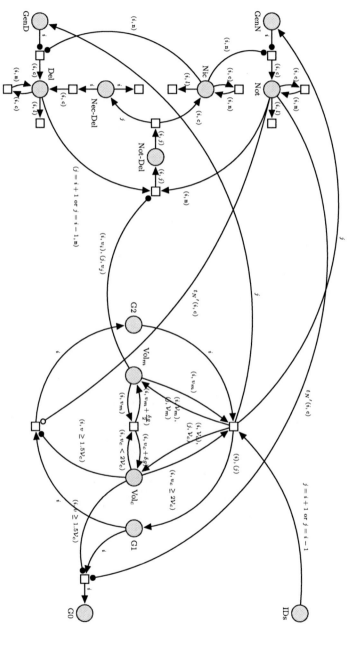

Figure 4.7: Colored Petri net of upward and downward causation processes. Place nodes Not, Nic, and Del are of type $N \times \{m, c, n\}$, Not-Del of type $N \times N$, Vol_c and Vol_m of type $N \times \mathbb{R}$, and the remaining places are of type N. Transition labels are omitted.

edges of the transition that describes Delta/Notch binding, then require the existence of tokens not only having adjacent index values but also showing the same cellular location of the membrane compartment in order to enable the transition. This way, the number of place nodes can be further reduced compared to Figure 4.6. To describe downward causation, the binding transition is connected a further pre-place, i.e. the membrane volume Vol_m. Unlike the extended Petri net model described in Chapter 4.1.2, here compartment volumes are represented by real-valued variables (tokens of type $\mathbb{N} \times \mathbb{R}$). The values of cells i and j are bound to the variables v_i and v_j respectively, which can then be used to dynamically adjust the firing rate of the transition (not shown in Figure 4.7).

Variables in a colored Petri net can also be used within expressions assigned to outgoing edges of a transition. Thereby, attribute values of produced tokens may depend on previously bound values of pre-place tokens. For example, the cell growth transition in Figure 4.7 computes the new values of the membrane and cytoplasm volumes based on the variables v_m and v_c that have been previously bound. Please notice the place node labeled with "IDs", whose initial marking determines the potential index values j of newly created cells.

4.2.2 Attributed π-Calculus

John et al. (2008, 2010) have introduced an attributed language as an extension of the π-calculus. The *attributed π-calculus* allows for defining attributed processes in order to describe different states or properties of biological processes, e.g., that are taking place within different cellular compartments. Moreover, by using functional λ-calculus expressions, the synchronization of processes may depend on their attribute values, which permits to describe complex rate laws and to constrain reactions flexibly. A stochastic semantics in terms of CTMCs allows for an application in systems biology.

To show the general idea of attributes and constraints in the realm of the π-calculus, Table 4.3 provides a set of process definitions that are describing compartmental translocation as well as binding processes of Delta and Notch. Each process is attributed by an $id \in \mathbb{N}$, which denotes a cell's index position in one-dimensional space. In addition, *Not*, *Del*, and *Nic* processes are also

Table 4.3: Attributed π-calculus model of Delta/Notch complexation and translocation. Expressions $e_1 \Rightarrow e_2$ are syntactic sugar to describe conditions of type: if e_1 then e_2 else 0. Similar to the simplification shown in Section 4.1.3, the $\tau[e]$ notation is used to avoid the definition of "dummy" co-processes for describing unimolecular reactions.

$Not(id, loc)$	::=	$bind[\lambda i.\lambda l.loc = \text{`mem'} \wedge l = \text{`mem'} \wedge ((id = i + 1) \vee (id = i - 1)) \Rightarrow k_5$
		$]?(cleavage).cleavage[k_6]!.Nic(id, \text{`cyt'}) +$
		$\tau[loc = \text{`cyt'} \Rightarrow k_3].Not(id, \text{`mem'})$
$Del(id, loc)$	::=	$(\nu\ cleavage)\ bind[id\ loc]!(cleavage).cleavage[\lambda k.k]?.NecDel(id) +$
		$\tau[loc = \text{`cyt'} \Rightarrow k_4].Del(id, \text{`mem'})$
$Nic(id, loc)$	::=	$\tau[loc = \text{`cyt'} \Rightarrow k_7].Nic(id, \text{`nuc'})$
$NecDel(id)$	::=	$\tau[k_8].Del(id, \text{`cyt'})$

attributed by a location $loc \in \{\text{`mem'}, \text{`cyt'}, \text{`nuc'}\}$ describing the cellular compartment within which the process is operating.

The major difference to the original π-calculus is that the synchronization of processes in the attributed π-calculus not only depends on the names of communication channels, but also on functional expressions that may operate on attribute values and – in the stochastic semantics – evaluate to positive real numbers, denoting the stochastic rates of interactions. For example, for successful communication between two individual Del and Not processes, the former process first needs to provide the values of its attributes id and loc, which is specified by $bind[id\ loc]!$. A process Not receives these values and checks for the correct compartmental location (i.e., the membrane, denoted by `mem`) and whether both processes have adjacent cell indices (i.e., $(id = i + 1) \vee (id = i - 1)$, where id denotes the index of Not and i the index of Del). Consequently, binding between both processes with rate k_5 is permitted only if these constraints are fulfilled. The expression of type $e_1 \Rightarrow e_2$ thereby is syntactic sugar replacing the frequently applied conditional expression if e_1 then e_2 else 0, where expression e_1 needs to evaluate to the Boolean value "true" as a prerequisite for communication. Once the expression evaluates successfully to a positive real number, i.e., parameter k_5, communication is enabled and *Del* transmits the private channel *cleavage* to the receiving process *Not*, over which a subsequent interaction of both processes occurs with rate k_6. Finally, *Not* will proceed with a process *Nic*(*id*, 'cyt') and *Del* with a process *NecDel*(*id*). Unimolecular translocations from one compartment into another are described by $\tau[\ldots]$, which is again syntactic sugar preventing from defining "dummy" co-processes. For example, the process *Nic*(*id*, *loc*) in Table 4.3 can be replaced by the following two processes:

$$Nic(id, loc) \quad ::= \quad transloc[\lambda i.\lambda l.\text{if } i = id \wedge l = loc \wedge loc = \text{'cyt' then } k_7 \text{ else } 0$$
$$]?.Nic(id, \text{'nuc'})$$

$$Nic_{co}(id, loc) \quad ::= \quad transloc[id\ loc]!.Nic_{co}(id, loc)$$

As complex, attribute-dependent expressions may determine the rate of communication, the attributed π-calculus permits the description of dynamic processes with rather flexible kinetic laws. For example, the processes *GenN*(*id*) and *GenD*(*id*) in Table 4.4 describe dynamically regulated gene expression of Delta and Notch in compliance with the description in Chapter 3.1.3. However, this requires information about the total amount of Nicd within the nucleus of a given cell. Therefore, *GenN*(*id*) and *GenD*(*id*) communicate via the *genexpr* channel with a process *TotalNic_n*(*id*, *count*), which is attributed with the nuclear Nicd count of a cell with index *id*. As the total count may change due to individual translocation and degradation processes of *Nic*(*id*, *loc*), the population-based process *TotalNic_n*(*id*, *count*) needs to be immediately updated once an according event takes place. This is achieved by prioritized communication steps with infinitely large rates, which are performed before any normal (i.e., non-prioritized) interaction will be selected. For example, the total amount of nuclear Nicd may change either due to translocation from the cytoplasm to the nucleus or due to degradation. Therefore, synchronization between *Nic*(*id*, *loc*) and *TotalNic_n*(*id*, *count*) processes via channel *updnic* is defined in order to update the population-based *count* attribute of *TotalNic_n*(*id*, *count*):

$$TotalNic_n(id, count) \quad ::= \quad updnic[\lambda i.i = id \Rightarrow \infty]?(\delta).TotalNic_n(id, count + \delta) \ + \ \ldots$$

$$Nic(id, loc) \quad ::= \quad \tau[loc = \text{'cyt'} \Rightarrow k_7].updnic[id]!(+1).Nic(id, \text{'nuc'}) \ +$$
$$\tau[loc = \text{'nuc'} \Rightarrow k_{deg}].updnic[id]!(-1).\mathbf{0} \ + \ \ldots$$

Table 4.4: Attributed π-calculus model of the Notch signaling example. A detailed explanation of the dynamic processes is given in the text. Cells are assumed to divide into one direction only, i.e., the id of a newly instantiated daughter cells is incremented by 1. Also, like in previous examples, newly instantiated cells are assumed to not containing any Notch nor Delta molecules.

$Cell(id, vol_m, vol_c, ph)$ ::= $\tau[vol_c < 2\mathcal{V}_c \Rightarrow k_g \cdot vol_c].Cell(id, vol_m + \frac{\delta_v}{2}, vol_c + \delta_v, ph) +$
$chkpt[\lambda i.\lambda n.id = i \wedge ph = \text{`g1'} \wedge vol_c \geq 1.5\mathcal{V}_c \wedge n < t_N \Rightarrow \infty]?.Cell(id, vol_m, vol_c, \text{`g2'}) +$
$chkpt[\lambda i.\lambda n.id = i \wedge ph = \text{`g1'} \wedge vol_c \geq 1.5\mathcal{V}_c \wedge n \geq t_N \Rightarrow \infty]?.Cell(id, vol_m, vol_c, \text{`g0'}) +$
$\tau[ph = \text{`g2'} \wedge vol_c \geq 2\mathcal{V}_c \Rightarrow \infty].(Cell(id, \mathcal{V}_m, \mathcal{V}_c, \text{`g1'}) \mid Cell(id + 1, \mathcal{V}_m, \mathcal{V}_c, \text{`g1'}) \mid TotalNot_c(id + 1, 0) \mid$
$\qquad TotalNic_n(id + 1, 0) \mid GenN(id + 1) \mid GenD(id + 1))$

$TotalNot_c(id, count)$ ::= $updnot[\lambda i.i = id \Rightarrow \infty]?(\delta).TotalNot_c(id, count + \delta) + chkpt[id\ count]!.TotalNot_c(id, count)$

$TotalNic_n(id, count)$ ::= $updnic[\lambda i.i = id \Rightarrow \infty]?(\delta).TotalNic_n(id, count + \delta) + genexpr[id\ count]!.TotalNic_n(id, count)$

$GenN(id)$ ::= $(\tau[k_1^*] + genexpr[\lambda i.\lambda c.i = id \Rightarrow k_1 \frac{c^h}{K^h + c^h}]?).updnic[id]!(+1).(GenN(id) \mid Not(id, \text{`cyt'}))$

$GenD(id)$ ::= $genexpr[\lambda i.\lambda c.i = id \Rightarrow k_2(1 - \frac{c^h}{K^h + c^h})]?.(GenD(id) \mid Del(id, \text{`cyt'}))$

$Not(id, loc)$ ::= $bind[\lambda i.\lambda l.loc = \text{`mem'} \wedge l = \text{`mem'} \wedge ((i = id + 1) \vee (i = id - 1)) \Rightarrow k_5]?(cleavage).cleavage[k_6]!.Nic(id, \text{`cyt'}) +$
$\tau[loc = \text{`cyt'} \Rightarrow k_3].updnot[id]!(-1).Not(id, \text{`mem'}) +$
$\tau[loc = \text{`cyt'} \Rightarrow k_{deg}].updnot[id]!(-1).0 + \tau[loc = \text{`mem'} \Rightarrow k_{deg}].0$

$Del(id, loc)$ ::= $(\nu\ cleavage)\ bind[id\ loc]!(cleavage).cleavage[\lambda k.k]?.NecDel(id) +$
$\tau[loc = \text{`cyt'} \Rightarrow k_4].Del(id, \text{`mem'}) +$
$\tau[loc = \text{`cyt'} \Rightarrow k_{deg}].0$

$Nic(id, loc)$ ::= $\tau[loc = \text{`cyt'} \Rightarrow k_7].updnic[id]!(+1).Nic(id, \text{`nuc'}) +$
$\tau[loc = \text{`nuc'} \Rightarrow k_{deg}].updnic[id]!(-1).0 + \tau[loc = \text{`cyt'} \Rightarrow k_{deg}].0$

$NecDel(id)$ ::= $\tau[k_8].Del(id, \text{`cyt'}) + \tau[k_{deg}].0$

79

Synchronization via the *updnic* channel evaluates to an ∞ rate, if both processes have the same *id*. The *Nic* process then transmits the change in the amount of nuclear Nicd, i.e., $+1$ in the case of a translocation from the cytoplasm into the nucleus and -1 in the case of degradation. $TotalNic_n$ receives this value and proceeds with an attributed process storing the updated information about the total count. This information is then used to determine gene expression rates with Hill-type kinetics, as has been explained above.

In principle, a combination of individual-based and population-based processes would make it also possible to describe catalyzed cleavage of activated Notch with Michaelis-Menten kinetics. However, here we need to distinguish between two different total count variables per cell, as each cell may have two neighbors and thus also two different variants of Notch-Delta complexes may exist. To keep the model simple, Table 4.4 therefore does not include these dynamics, but shows another population-based process, namely the amount of cytoplasmic Notch. This information is used to describe the upward causation at the G_1 checkpoint of the cell cycle, where the excess of a threshold value t_N leads to cell cycle arrest in phase G_0 rather than the continuation with cell cycle phase G_2.

Although the attributed π-calculus allows to model the dynamic change of compartment volumes (cf. the *Cell* process in Table 4.4) and we have seen how to integrate such global information with individual process dynamics, describing volume-dependent rates of bimolecular binding reactions is difficult if not impossible for the Notch signaling example, which has different reasons. Taking global information into account similar to the previously described gene expression processes is not possible due to the restriction to binary interactions. Even without taking the volume as a side-condition into account in order to describe downward causation, binding between Delta and Notch requires already an interaction between two processes. Providing information about the volume of the membrane compartments of both adjacent cells would therefore require to define a synchronous interaction between four rather than two processes, which is impossible in π-calculus-based languages. One could circumvent this problem by additionally attributing Notch and Delta processes with the volume of the compartment they reside in. However, unlike using prioritized interactions to aggregate individual state changes in a many-to-one manner, this would require to instantaneously and simultaneously update *all* individual processes of a cell once its compartment volume changes, i.e., one-to-many, which can hardly be achieved.

The *imperative π-calculus* (John et al., 2009; John, 2010) – a formal language that is obtained from and thus shares many features with the attributed π-calculus – offers a solution for modeling such processes in dependence on global side-conditions. The imperative π-calculus allows defining global variables that can be accessed by individual processes (no example shown). This way, global stores can be used to describe reactions with Michaelis-Menten kinetics (Mazemondet et al., 2009) or dynamic compartment volumes (John et al., 2009). However, in the case of our multicellular example model this method does not denote a viable solution, as it would require the definition of a global variable for the volume of each distinct cell, which is impossible due to the dynamic instantiation of new cells during runtime. Therefore, the only potential alternative for describing the volume-dependent binding between Delta and Notch in the attributed π-calculus seems to be a predominantly population-based model, where each molecular species is described by an attribute of a few high-level processes (cf. John et al., 2010). However, this in turn counteracts the complexity reduction, as it prevents from using features like private channel names for restricting the communication scope.

4.2.3 Rule-based Approaches

Rule-based modeling denotes the usage of a specific class of modeling languages, in which dynamics are described by a set of rules whose execution (also application or firing) determine state changes of the model. Thereby, each rule comprises certain conditions that need to be fulfilled in order to be executed (Hayes-Roth, 1985). From a conceptual simulation strategy point of view, rule-based modeling focuses on reactions rather than objects and follows the activity scanning approach (cf. page 38f).

Rule-based models may be specified by rather narrative descriptions, like the qualitative ecological models by Starfield (1990), for instance. However, in the context of quantitative systems biology, the syntax of rule-based languages typically follows the notation of chemical reaction equations or rather similar representations, i.e., the left-hand side of a rule defines the condition for rule application while the right-hand side specifies the action to be performed once the rule fires. Just to mention some examples, BioNetGen Language (Faeder et al., 2005, 2009), κ-calculus (Danos et al., 2007a, 2009), BIOCHAM (Chabrier-Rivier et al., 2005; Fages and Soliman, 2008), and LBS (Pedersen and Plotkin, 2010) are rule-based languages that fall into this category. These languages – like many other rule-based approaches – can be also classified as *attributed* languages, since rules are applied to structured (i.e., attributed) model entities (sometimes termed agents) with possibly various states. By employing conditional patterns, *rules* become *rule schemata*, which may be instantiated in different contexts. Therefore, rule-based modeling has been found to be a powerful tool for managing combinatorial explosion and it enables a concise and compact description of biochemical and cell biological models (Hlavacek et al., 2006; Blinov and Moraru, 2012).

Let us take an example in the BioNetGen Language (BNGL). The basic model entities in BNGL are *molecules* that may be attributed by so called *components*. The BNGL model in Figure 4.8 comprises a "Not" and a "Del" molecule type, both consisting of multiple components that need to be specified by a name and a set of potential values (states) of each component. Consequently, the restriction to a finite value set hampers the description of newly generated cells, for instance, as the identifier ("ID") of each potentially appearing cell needs to be specified in advance. However, on the other hand, only a limited number of intracellular compartmental locations ("LOC") exist where Notch and Delta may reside in, namely the membrane ("mem"), cytoplasm ("cyt"), and the nucleus ("nuc"). Please note, the "memnecd" location of "Del" denotes a Delta ligand molecule within the membrane compartment right after the cleavage of a bound Notch receptor. To distinguish between the mature Notch receptor (consisting of both the extracellular as well as intracellular domain) and Nicd, "Not" molecules have a component "D" with either state "ei" or "i". In addition, "Not" and "Del" molecules both have another component "BS" having no defined state but denoting the binding site between Delta ligand and Notch receptor. Binding between molecules can be described

```
begin molecule types
  GenN(ID~1~2)
  GenD(ID~1~2)
  Not(ID~1~2,LOC~cyt~mem~nuc,D~ei~i,BS)
  Del(ID~1~2,LOC~cyt~mem~memnecd,BS)
  Trash()
end molecule types

begin observables
  Molecules Nic_n1 Not(ID~1,LOC~nuc) # amount of nuclear Nicd in cell1
  Molecules Nic_n2 Not(ID~2,LOC~nuc) # amount of nuclear Nicd in cell2
end observables

begin functions
  Theta1() = (Nic_n1^h) / (K^h + Nic_n1^h)
  Theta2() = (Nic_n2^h) / (K^h + Nic_n2^h)
end functions

begin reaction rules
  # gene expression
  GenN(ID~1) -> GenN(ID~1) + Not(ID~1,LOC~cyt,D~ei,BS) k1prime + k1*Theta1()
  GenN(ID~2) -> GenN(ID~2) + Not(ID~2,LOC~cyt,D~ei,BS) k1prime + k1*Theta2()
  GenD(ID~1) -> GenD(ID~1) + Del(ID~1,LOC~cyt,BS)       k2 * (1 - Theta1())
  GenD(ID~2) -> GenD(ID~2) + Del(ID~2,LOC~cyt,BS)       k2 * (1 - Theta2())

  # protein translocation
  Not(LOC~cyt,D~ei) -> Not(LOC~mem,D~ei) k3
  Del(LOC~cyt)      -> Del(LOC~mem)      k4
  Not(LOC~cyt,D~i)  -> Not(LOC~nuc,D~i)  k7
  Del(LOC~memnecd)  -> Del(LOC~cyt)      k8

  # binding between Delta and Notch
  Not(ID~1,LOC~mem,BS) + Del(ID~2,LOC~mem,BS) -> \
    Not(ID~1,LOC~mem,BS!1).Del(ID~2,LOC~memnecd,BS!1) k5
  Del(ID~1,LOC~mem,BS) + Not(ID~2,LOC~mem,BS) -> \
    Del(ID~1,LOC~memnecd,BS!1).Not(ID~2,LOC~mem,BS!1) k5

  # catalyzed cleavage of Notch
  Not(LOC~mem,D~ei,BS!1).Del(BS!1) -> \
    Not(LOC~cyt,D~i,BS) + Delta(BS) MM(kcat,Km)

  # protein degradation
  Not(BS) -> Trash() kdeg
  Del(BS) -> Trash() kdeg
end reaction rules
```

Figure 4.8: BNGL model of the Notch signaling example. The model describes biochemical dynamics within and between two cells only. *Parameters, seed species,* and *actions* blocks are omitted.

by using bond labels ("!1") and the "." symbol to indicate a molecular complex:

```
Not(BS) + Del(BS) -> Not(BS!1).Del(BS!1) k5
```

In Figure 4.8, a binding reaction is further constrained to molecules that are located within the membrane compartment ("LOC~mem") of adjacent cells ("ID~1" and "ID~2"):

```
Not(ID~1,LOC~mem,BS) + Del(ID~2,LOC~mem,BS) -> ...
```

and

```
Del(ID~1,LOC~mem,BS) + Not(ID~2,LOC~mem,BS) -> ...
```

As BNGL – unlike the attributed π-calculus or colored Petri nets – does not support the specification of attribute patterns by utilizing mathematical expressions, we need to encode the neighborhood by explicitly inserting defined cell "ID"s of each reactant molecule. Hence, two distinct rules need to be defined in order to model Delta/Notch binding in a system comprising of two cells, which does not denote any complexity reduction compared to ODEs or conventional Petri net models, for instance. However, the employment of more expressive reaction constraints in a rule-based formalism is also possible, as has been successfully shown by Kuttler et al. (2010) and John et al. (2011). Also, even without such capability the combinatorial complexity can be reduced. As the "ID" of "Not" and "Del" will remain unchanged – so that the molecule's cellular locality is preserved during the bound state – and BNGL follows the "*don't care, don't write*" mantra, the subsequent cleavage of activated Notch requires only a single rule schema, no matter how many cells do exist.

Similarly, the two protein degradation rules "Not(BS) -> Trash() kdeg" and "Del(BS) -> Trash() kdeg" encode for the degradation of any "Not" and "Del" molecule as long as having an unbound binding site "BS". Therefore, together both degradation rules encode for a set of 14 elementary reactions (7 per cell, cf. Table 3.1 on page 45) in the two-cell model of Figure 4.8. Please notice, "Trash" denotes a dummy molecule, which is necessary since BNGL requires one product species on the right-hand side of each rule at least.

```
begin observables
  Molecules Vol_m1 Vol(ID~1,LOC~mem)        # membrane volume of cell1
  Molecules Vol_m2 Vol(ID~2,LOC~mem)        # membrane volume of cell2
  Molecules Vol_c1 Vol(ID~1,LOC~cyt)        # cytoplasm volume of cell1
  Molecules Not_c1 Not(ID~1,LOC~cyt,D~ei)   # cytoplasmic Notch amount in cell1
  ...
end observables

begin reaction rules
  # binding between Delta and Notch
  Not(ID~1,LOC~mem,BS) + Del(ID~2,LOC~mem,BS) -> \
    Not(ID~1,LOC~mem,BS!1).Del(ID~2,LOC~memnecd,BS!1) k5 / (Vol_m1 + Vol_m2)
  ...

  # cell cycle phase transition from G1 to G2
  Cell(ID~1,PHASE~G1) -> Cell(ID~1,PHASE~G2) \
    if(Vol_c1 >= 1.5*Vc, if(Not_c1 < tN, kG2, 0), 0)
  ...

  # cell cycle arrest (transition from G1 to G0)
  Cell(ID~1,PHASE~G1) -> Cell(ID~1,PHASE~G0) \
    if(Vol_c1 >= 1.5*Vc, if(Not_c1 >= tN, kG0, 0), 0)
  ...
end reaction rules
```

Figure 4.9: BNGL rules describing upward and downward causation. Conditional expressions are defined as if(boolean condition, true clause, false clause).

The recent BNGL specification[2] also supports arbitrary rate kinetics (in past versions, reactions were restricted to follow the law of mass action, Michaelis-Menten kinetics, or a saturation rate law). Arbitrary kinetics can be specified by user-defined functions, which may depend on global observables (Sneddon et al., 2011). For example, in order to model the Hill-type gene expression rates, two observables "Nicd_nuc1" and "Nicd_nuc2" are defined in Figure 4.8, which hold the current amounts of nuclear Nicd in each of both cells and determine the functional values of "Theta1" and "Theta2" that are used to describe the rate law of according gene expression rules. In the same way, volume-dependent downward causation can be described (Figure 4.9). However, similar to the Petri net in Figure 4.4, compartment volumes need to be described by the same kind of model entity like other state variables, i.e., volume size can

---

[2]see http://bionetgen.org/index.php/Release_Notes#version_2.2.0_stable

be described in terms of *molecule* count only. As a consequence, the quantities of compartment volumes are represented by discrete integers rather than real numbers and therefore require a recalculation and normalization of certain parameters. Moreover, this also hampers the description of abrupt volume changes, e.g., due to cell division.

Global observables can be also used to describe cell cycle phase transitions of the example model that are constrained by upward causation (second and third rule in Figure 4.9). Conditional expressions evaluate the according time-dependent variables such that, e.g., a transition from G_1 to G_2 phase is prevented (0-rate) if the volume of the cytoplasm is lower than 150% of the initial value ("$1.5*Vc$") or the amount of cytoplasmic Notch exceeds a certain threshold value ("tN"). Due to the need for global observables to access the total amounts of certain molecular species, it is not possible to describe the cell cycle transition rules in terms of rule schemata, i.e., to ignore the "ID" attribute of "$Cell$".

4.3 Concluding Remarks

In this chapter, an overview of diverse flat modeling approaches has been given and their support or suitability for describing different multilevel aspects has been discussed with the help of the Notch signaling example from Chapter 3. Table 4.5 tries to summarize the results, however, please notice that the assessment may be rather imprecise and subjective.

By definition, flat approaches do not support any hierarchical structuring of models. Therefore, multiple (nested) levels can be described only implicitly, which is typically realized by an appropriate naming of different state variables, e.g., $Prot_{cyt}$ and $Prot_{nuc}$ to distinguish between the cytoplasmic and nuclear location of a certain kind of protein. The size of model descriptions, e.g., the number of species, equations, or graph nodes, may thereby easily outreach manageable dimensions (cf. the Petri net in Figure 4.2 that describes a rather simple reaction network within and between merely three cells).

Private channel names in the π-calculus can be used to describe compartmentalized dynamics by restricting the communication scope of processes.

Table 4.5: Summary of the expressivity of diverse flat modeling approaches with respect to describing certain multilevel aspects. Good and moderate support is denoted by "+" and "o" respectively, while "−" denotes no or only poor support for modeling the according aspects.

	Hierarchical model structure	Dynamic model structure	State & behavior at diff. levels	Arbitrary rate laws & constraints	Upward & downward causation	Space beyond compartments
Ordinary differential equations	−	−	−	o	o	o
Extended stochastic Petri nets	−	−	−	+	o	−
Stochastic π-calculus	−	o	−	−	−	o
Colored stochastic Petri nets	−	−	o	+	o	+
Attributed π-calculus	−	o	o	o	o	+
BioNetGen Language	−	o	o	+	o	−

Thereby, the number of process definitions may be effectively reduced. Another approach for describing compartmentalized dynamics more conveniently is to use attributes like in colored Petri nets or most rule-based approaches, where different locations may be modeled by assigning according attribute values to model entities. This way, molecules within different compartments can be clearly distinguished from each other, while models can be kept small due to generalized transitions or rule schemata. However, structural relationships between different compartments still remain implicit. In other words, there is no way to explicitly describe hierarchical nesting, possibly having a negative effect on the understandability of multilevel models.

Describing phenomena like cell division requires adding state variables dynamically, which is not supported by ODEs and Petri nets. Adding new cells dynamically is also difficult or even impossible to describe in the π-calculus and

BNGL, however, these languages allow for describing dynamic model structures originating from molecular bindings. Please note, although colored Petri nets do not allow for changing model structures dynamically, the process of cell division can be mimicked to some degree by its colored token sets, similar to the approach employed by the attributed π-calculus.

Equally important, diverse examples indicate that supporting dynamic processes to be flexibly constrained facilitates the description of certain multilevel aspects. Thereby, constraints may often be based on individual (local) attributes, e.g., the translocation of a molecule from one compartment into another is constrained by its current location and mathematical expressions facilitate the description of spatial relationships based on corresponding attributes. In addition, also information of an *extended scope* may be of importance to model approximate reaction kinetics beyond the usual law of mass action and particularly to describe interlevel causation. For instance, the aggregated amount of cytoplasmic Notch constrains cell cycle phase transitions (upward causation) and a bimolecular reaction rate depends on the surrounding compartment volume (downward causation).

Due to the lack of structuring elements, state variables denoting such high-level information are oftentimes described in the same way like any other model entity and therefore may misleadingly appear to be qualitatively equivalent, e.g., molecules and compartment volumes in a Petri net that are described by the same kind of place nodes. Also, in this case the constraining information denotes an additional side-condition that increases the number of entities taking part in a dynamic process. Therefore, termolecular reactions (or reactions comprising of even higher molecularities) may need to be defined, which is impossible in some formalisms, e.g., the π-calculus is restricted to binary interactions.

In an attributed language, such high-level information may be assigned to defined "high-level" entities. For example, in the attributed π-calculus model in Table 4.4 the membrane and cytoplasm volumes as well as the phase of the cell cycle are attributes of the *Cell* process rather than described by own distinct processes. This way, not only a basic structuring of the model has been realized (compartment volumes and the cell cycle phase are properties of the same entity), but also multiple side-conditions can be taken into account without increasing the number of interacting entities. Different types of attributes thereby facilitate an appropriate representation of different kinds of information, e.g, real numbers or strings of characters.

Besides attributing certain model entities, another approach for constraining dynamic processes based on high-level information is to define according global variables that can be accessed everywhere, as it can be done in the imperative π-calculus and BNGL, for instance. In a single-level model this seems to be an effective approach, however, modeling multilevel systems comprising of many compartments and cells requires the definition of numerous global variables denoting similar information in different contexts. The number of global variables to be defined may therefore become rather large. Moreover, adding global variables dynamically is typically not supported, which again hampers the description of dynamic structure models.

Chapter 5

Hierarchically Structured Models

In the previous chapter, an overview of diverse flat modeling approaches has been given. It has been shown that certain formalisms are better suited for describing multilevel behavior than others. However, a common limitation is that – by definition – each flat modeling approach allows to describe nested hierarchies only implicitly. Different modeling approaches with an explicit support for hierarchies and their suitability for describing biological systems at multiple levels are therefore discussed in this chapter. Thereby, the Notch signaling model presented in Chapter 3 again serves as a running example to illustrate how different aspects of biological multilevelness can be expressed.

At first, different hierarchical extensions of originally flat formalisms are discussed, i.e., hierarchical Petri nets, BioAmbients as an extension of the π-calculus, and an extension of the rule-based BioNetGen Language supporting nested compartments. Subsequently, inherently hierarchical modeling approaches like Statecharts and DEVS are in the focus of Chapter 5.2. At the end of this chapter, a short summary of the different approaches will be given.

5.1 Hierarchical Extensions of Flat Formalisms

5.1.1 Hierarchical Petri Nets

To facilitate the construction of large Petri nets, Huber et al. (1991) proposed to structure P/T nets hierarchically by means of interrelated subnets. The proposed constructs thereby do not extend the expressive power of Petri nets,

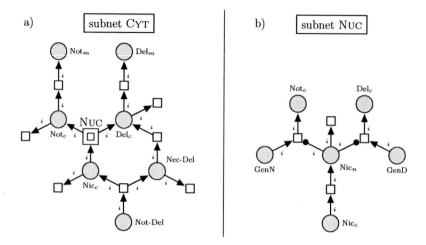

Figure 5.1: Subnets of a hierarchical colored Petri net of the Notch signaling example. (a) Subnet CYT representing the dynamic processes taking place within the cytoplasm. (b) Subnet NUC of nuclear processes. All place nodes in both subnets are of type $\mathbb{N}$. Transition labels are omitted.

but provide a formal framework for composition and explicit structuring of Petri nets in a hierarchical manner. The main concepts that distinguish hierarchical from conventional (non-hierarchical) Petri nets are *substitution nodes* and *fusion sets*.

Substitution allows replacing entire subnets consisting of numerous places and transitions by so-called *macro nodes*. Thereby, a system may be described in a component-based manner and in terms of different levels of detail. For example, the macro transition NUC in Figure 5.1a denotes a rather abstract representation of the nuclear dynamics of our Notch signaling example model, namely the consumption of cytoplasmic Nicd (Nic_c) and the production of Notch receptors (Not_c) as well as Delta ligands (Del_c) by the nucleus. The actual detailed dynamics, like the translocation of Nicd from the cytoplasm into the nucleus and the synthesis of Notch and Delta in dependence on the nuclear amount of Nicd, are described by a distinct subnet NUC (Figure 5.1b) substituting the macro transition in subnet CYT (Figure 5.1a). Thereby, the interfaces for composition are denoted by identically named place nodes in

CYT and NUC (i.e., Nic_c, Not_c, and Del_c) that are semantically merged. Subnet CYT, in turn, denotes merely a component of a larger Petri net shown in Figure 5.2. This high-level Petri net looks pretty much similar and is completely equivalent to the colored Petri net depicted in Figure 4.7 on page 75. However, besides the substitution transition (CYT), it makes additionally use of another concept of hierarchical Petri nets, i.e., fusion sets.

A fusion set defines a set of nodes that are functionally identical. That means, each action at one of the nodes instantaneously also takes place at all other nodes of the set and thus the nodes of a fusion set can be considered to be copies of each other. Three different sets of fusion places are defined in Figure 5.2 to describe certain interlevel dynamics influenced by or influencing the markings of place nodes within subnets CYT and NUC respectively. Without the capability to define fusion places it would be impossible to describe the multilevel model in terms of distinct subnets, as, e.g., the amount of cytoplasmic Notch (Not_c) needs to be taken into account for modeling certain high-level processes of the cell cycle but its own dynamics is described at a lower hierarchy level, i.e., within subnet CYT. Therefore, fusion sets allow for describing upward and downward causation across different levels of organization.

Although subnets may be called more than once within a hierarchical Petri net, a dynamic instantiation during runtime is not possible and thus the structure of such a model is fixed. This is different in *object-oriented* Petri nets such as *Mobile Nets* (Busi, 1999) or *Reference Nets* (Kummer, 2001), which also address the need for describing systems at multiple levels by following a *"nets-within-nets"* paradigm, i.e., they support to have nets as tokens of other nets (see also Lomazova, 2000, 2001; Valk, 2004; Miyamoto and Kumagai, 2005; Köhler-Bußmeier, 2009). Thereby, migration of entire subnets within high-level structures becomes possible, facilitating the description of dynamic variable structure models.

Communication between different levels of an object-oriented Petri net is typically realized by *synchronous channels* that enable a synchronized firing of two transitions assigned with the same channel name (Christensen and Damgaard Hansen, 1994). However, since this method relies on binary interac-

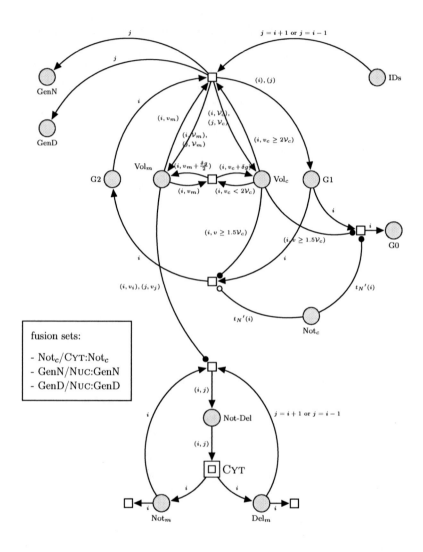

Figure 5.2: Hierarchical colored Petri net of the Notch signaling example. Place node Not-Del is of type $\mathbb{N} \times \mathbb{N}$, Vol_c and Vol_m are of type $\mathbb{N} \times \mathbb{R}$, and the remaining places are of type $\mathbb{N}$. Transition labels are omitted.

tions and is therefore quite similar to process communication in the π-calculus, it may be difficult to take additional side effects into account. Also, object-oriented Petri nets are typically executed in equidistant time steps rather than discrete events in continuous time, which hampers modeling of biochemical systems.

5.1.2 BioAmbients

In flat formalisms, the representation of different locations like different cells or cellular compartments often requires to define different "variants" of an entity by introducing distinct entity names, one for each location. By contrast, the π-calculus allows to describe different compartments also by restricting communication scopes achieved by private channel names and according process parameterization, as has been shown in Chapter 4.1.3 (see also Regev, 2002). However, both methods – i.e., defining distinct entity names as well as using private communication channels – are often impractical as they allow, for example, to describe model structures only implicitly. Moreover, the number of entities to be defined may be prohibitive and modeling dynamic structures like compartment fusion or migration is impossible or requires complicated workarounds for propagating private channel names. Therefore, Regev et al. (2004) have presented BioAmbients as a solution to overcome these problems.

BioAmbients extends the π-calculus by additional means for structuring models explicitly and thereby restricting the communication scopes of processes. Based on the calculus of mobile ambients (Cardelli and Gordon, 1998), the extension allows for wrapping processes P by so-called *ambients*, which is written as $[P]$. Ambients may have an optional name for annotation purposes and they may be arbitrarily nested, which is of importance to appropriately reflect the hierarchical nature of biological compartments. For instance, the intracellular compartmentalization of the Notch example, where each compartment may contain different kinds of protein molecules, can be described by nesting and parallel composition of different sub-ambients (membrane, cytoplasm, nucleus) and protein-representing processes (*Del*, *Not*, *Nic*):

$$\text{membrane}[\,Del\,|\,Not\,|\,\ldots\,|\,\text{cytoplasm}[\,Del\,|\,Not\,|\,Nic\,|\,\ldots\,|\,\text{nucleus}[\,Nic\,|\,\ldots\,]\,]\,]$$

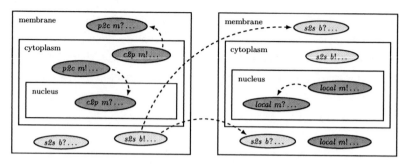

Figure 5.3: Restricted communication scope within nested ambients. Processes that may interact with each other are connected by dashed arrows.

The nesting of ambients may be dynamically changed by applying so called *capabilities*, for which pairwise synchronization via identical channel names and direct adjacency of ambients is required. Different capability pairs thereby describe different variable structure operations: *enter/accept* allows an ambient to enter another ambient at the same hierarchy level, *exit/expel* describes the opposite mechanism, and *merge+/merge−* allows two ambients located at the same level to merge. In all cases, the content of ambients will be entirely moved or merged in conjunction, e.g., merging two ambients [*merge+ m.A*] and [*merge− m.A | B*] by synchronizing via channel *m* results in a single ambient [*A | A | B*].

Besides capabilities for changing the nested structure, the ambient hierarchy also restricts communication between processes, i.e., two processes may communicate with each other only if they are in proximity. Similar to different capability pairs, process communication is additionally constrained by pairs of compatible communication directions: *local/local* for process communication within the same ambient, *p2c/c2p* for communication from a parent to its child ambient, *c2p/p2c* for the opposite direction, and *s2s/s2s* describes communication between two processes residing within different ambients that are direct siblings. Figure 5.3 illustrates such restricted communication scope of processes within hierarchically nested ambients. A structured model of the biochemical processes of the Notch signaling example that makes use of all different communication directions is shown in Table 5.1.

Table 5.1: BioAmbients model of biochemical Notch signaling processes. The model describes signaling between two cells. Synchronization rates are omitted.

Init	::=	membrane[*Mem* | cytoplasm[*Cyt* | nucleus[*Nuc* | *GenN* | *GenD*]]] |
		membrane[*Mem* | cytoplasm[*Cyt* | nucleus[*Nuc* | *GenN* | *GenD*]]]
Mem	::=	*p2c mnot?.(Mem | Not)* + *p2c mdel?.(Mem | Del)* + *local deg!.Mem*
Cyt	::=	*p2c mnot?.(Cyt | Not)* + *p2c mdel?.(Cyt | Del)* +
		c2p mnic?.(Cyt | Nic) + *c2p mdel?.(Cyt | Del)* + *local deg!.Cyt*
Nuc	::=	*c2p mnic?.(Nuc | Nic)* + *local basal?.Nuc* + *local deg!.Nuc*
GenN	::=	*local basal!.c2p mnot!.GenN* + *local act?{b}.local b!.c2p mnot!.GenN*
GenD	::=	*local basal!.c2p mdel!.GenD* + *local rep?{b}.local b!.GenD*
Nic	::=	*(ν b) local act!{b}.local b?.Nic* + *local rep!{b}.local b?.Nic* +
		p2c mnic!.0 + *local deg?.0*
Not	::=	*s2s bind?{b}.s2s b!.p2c mnic!.0* + *c2p mnot!.0* + *local deg?.0*
Del	::=	*(ν b) s2s bind!{b}.s2s b?.(local deg!.0* + *p2c mdel!.0)* +
		c2p mdel!.0 + *local deg?.0*

In this model, compartments are described as ambients (with names membrane, cytoplasm, and nucleus) and proteins as processes (*Nic, Not, Del*). Since capabilities like *enter* and *exit* can be used for changing the nested structure of entire ambients only, the translocation of protein molecules from one compartment into another is described by restricted interaction between protein processes and additional processes (*Mem, Cyt,* and *Nuc*) defined as communication partners within the respective compartment ambients. For example, translocation of cytoplasmic Notch to the membrane is described by an interaction between a *Not* and the *Mem* process over channel *mnot*, where due to the "child-to-parent" (*c2p/p2c*) relation communication is restricted to processes *Not* that are one level down the hierarchy compared to process *Mem*, i.e., to those that reside within the cytoplasm ambient:

$$Not \quad ::= \quad c2p \; mnot!.0 \; + \; \ldots$$
$$Mem \quad ::= \quad p2c \; mnot?.(Mem \,|\, Not) \; + \; \ldots$$

97

Please note, synchronization rates are omitted in the model for the sake of simplicity, although stochastic execution semantics for BioAmbients have been defined (Brodo et al., 2007; Phillips, 2009). These semantics, however, allow for standard mass-action kinetics only and therefore it is not possible to describe the dynamics of gene expression and cleavage of activated Notch as has been introduced in Chapter 3.1.3. Since ambients do not have a state and behavior of their own and the formalism supports only binary interactions like in the original π-calculus, it is also still impossible to model dynamic compartment volumes and according downward causation effects on Delta/Notch binding. The same holds true for cell cycle dynamics constrained by upward causation. Furthermore, BioAmbients also lacks the expressive power to constrain process communication based on more elaborate spatial relationships like the linear organization of multiple cells.

However, the BioAmbients calculus – like other related membrane-inspired languages such as Brane Calculi (Cardelli, 2005) and P systems (Păun and Rozenberg, 2002; Ardelean and Cavaliere, 2003; Spicher et al., 2008) – still denotes a significant step toward accessible multilevel modeling as it allows to describe nested model structures explicitly, restricts the communication scope of entities in dependence on these structures, and moreover also allows to change a model's structure dynamically.

5.1.3 Compartmental BNGL

The need for structuring models hierarchically has been also addressed in the realm of rule-based modeling, e.g., the bioκ-calculus (Laneve and Tarissan, 2008) that combines ideas of Brane Calculi (Cardelli, 2005) and the κ-calculus (Danos et al., 2007a), or the Calculus of Wrapped Compartments (Coppo et al., 2010b) as a variant of the Calculus of Looping Sequences (Barbuti et al., 2006, 2007). Another rule-based approach by Harris et al. (2009) – i.e., compartmental BNGL – shall be discussed in more detail here.

Compartmental BNGL (cBNGL) extends the originally flat BioNetGen Language by means of reaction compartments and membranes to explicitly describe the hierarchical topology of a cell and to take effects on reaction rates into account in dependence on the location of biochemical species. That means, in cBNGL bimolecular and higher-order reaction rates are automatically adjusted according to the compartment volume the reaction is taking place in. Thereby, the generality of rules for encoding different reactions may be increased. The localization of molecules also restricts the scope of rule application such that molecules in the same or adjacent compartments may interact with each other only.

A model's compartment topology is specified within the compartments block of a cBNGL model specification (see left panel on the top of Figure 5.4). Thereby, compartments may be defined either as three-dimensional or two-dimensional (i.e., membrane-like surfaces) compartments and each compartment has assigned a volume as well as surrounding parent compartment (except of the top-most compartment, which has no parent). Each molecule is localized within one of the specified compartments. Therefore, no additional component (i.e., attribute) needs to be defined in order to describe different locations of a molecule (cf. non-compartmental BNGL model in Figure 4.8, page 83). Different parts of a molecular complex may be localized within different compartments, which facilitates modeling of transmembrane and other membrane-anchored proteins consisting of multiple subdomains. This allows, for example, describing the Notch receptor in terms of three distinct parts, i.e., the intracellular domain Nicd, the extracellular domain Necd, and an intermediate transmembrane domain Nmd, each of which may be localized within a different compartment. Unfortunately, cBNGL does not support molecular species spanning multiple surface (i.e., membrane) compartments (Harris et al., 2009, Sect. 4.4), which hampers the description of intercellular Notch signaling. Therefore, the example model in Figure 5.4 describes the biochemical dynamics of Notch signaling within a single cell only.

By explicitly specifying the compartment name in postfix or prefix notation, transport rules may change the location of individual molecules and entire molecular complexes respectively. However, the compartment topology of a model is fixed, i.e., cBNGL does not provide means for dynamic structures. Also, besides the volume, it is not possible to equip compartments with an own state. Moreover, the volume of a compartment denotes merely a constant model parameter rather than a state variable that may change over

```
begin compartments                              begin species
  EXT 3 V_e     # extracellular space             DNA(gene~N)@NUC 1
  MEM 2 V_m EXT # cell membrane                    DNA(gene~D)@NUC 1
  CYT 3 V_c MEM # cytoplasm                       end species
  NUM 2 V_0 CYT # nuclear membrane
  NUC 3 V_n NUM # nuclear compartment
end compartments                                begin observables
                                                  # amount of nuclear Nicd
                                                  Molecules Nicd_n Nicd(m)@NUC
begin molecule types                            end observables
  DNA(gene~N~D)
  Nicd(m)
  Nmd(i,e)                                       begin functions
  Necd(m,d)                                        Theta() =
  Delta(e)                                           (Nicd_n^h) / (K^h + Nicd_n^h)
  Trash()                                         end functions
end molecule types
```

```
begin reaction rules
  # gene expression
  DNA(gene~N) -> DNA(gene~N) + @CYT:Nicd(m!1).Nmd(i!1,e!2).Necd(m!2,d) \
    k1prime + k1*Theta()
  DNA(gene~D) -> DNA(gene~D) + @CYT:Delta(e) k2 * (1 - Theta())

  # protein translocation and recycling
  Nmd(e!1)@CYT.Necd(m!1)@CYT -> Nmd(e!1)@MEM.Necd(m!1)@EXT k3
  Delta(e)@CYT -> Delta(e)@MEM k4
  Nicd(m)@CYT -> Nicd(m)@NUC k7
  Delta(e!1)@MEM.Necd(d!1,m) -> Delta(e)@CYT k8

  # receptor-ligand binding (misleadingly described by Notch and Delta
  # molecules being part of the same cell)
  Necd(m!+,d)@EXT + Delta(e)@MEM -> Necd(m!+,d!1)@EXT.Delta(e!1)@MEM k5

  # catalyzed cleavage of Notch
  Nicd(m!1).Nmd(i!1,e!2).Necd(m!2,d!3).Delta(e!3) -> \
    Nicd(m) + Necd(m,d!3).Delta(e!3) MM(kcat,Km)

  # protein degradation
  Nicd(m) -> Trash() kdeg
  Necd(m!+,d) -> Trash() kdeg
  Delta(e) -> Trash() kdeg
end reaction rules
```

Figure 5.4: Compartmental BNGL model of biochemical Notch signaling processes within a single cell. *Parameters* and *actions* blocks are omitted.

time. Therefore, modeling of downward causation and certain high-level aspects of the Notch signaling example is hampered, e.g., cell cycle dynamics in dependence on the cytoplasmic compartment volume. Upward causation in dependence on the cytoplasmic amount of Notch receptors could be described like in standard BNGL, i.e. by defining and accessing an according observable (cf. Figure 4.8). However, this would still require the definition of an own global observable as well as reaction rule for each modeled cell.

5.2 Inherently Hierarchical Approaches

5.2.1 Statecharts

The visual *Statecharts* formalism by David Harel (1987) has been developed to facilitate the description of *reactive systems*, i.e., of systems that react to external stimuli (Harel and Pnueli, 1985). A reactive system is characterized by different states reflecting different situations in the life of this system, during which it performs some action or waits for some event. Therefore, the semantics of Statecharts follows a discrete event-based approach, meaning that *"the behavior of a reactive system is described as a sequence of discrete events that cause abrupt changes (taking no time) in the state of the system, separated by intervals in which the system's state remains unchanged."* (Kesten and Pnueli, 1991, p. 592). Due to their clearly defined graphical representation and the discrete events semantics, Statecharts are prominent in the domain of model-based software engineering and have become an essential part of the standardized *Unified Modeling Language* (UML; Fowler, 2004). However, Statecharts have been also repeatedly applied to modeling biological systems, especially with a focus on multicellular models (see, e.g., Kam et al., 2001; Efroni et al., 2003; Amir-Kroll et al., 2008; Kugler et al., 2010; Fisher and Harel, 2010).

Statecharts are strongly inspired by conventional finite state machines, however, in addition they support hierarchical nesting and orthogonal (concurrent) states to structure a model in different ways. While hierarchy allows describing a system at different abstraction levels, i.e., refinements and generalizations, orthogonality helps to reduce combinatorial complexity by structuring the state of components into multiple concurrent sub-states. Transitions between different states are labeled as follows: "trigger[condition]/action", where each part of this labeling scheme is optional. Applications of transitions are triggered by the passage of time or by events generated as an action by other state transitions, which allows letting different parts of a model communicate with each other. In addition, conditions may constrain state transition firings based on other states.

Let us consider an example. The Statecharts model depicted in Figure 5.5 describes different states and state transitions of both Notch as well as Delta protein molecules in accordance with our recurring example model. Thereby, both proteins may either exist (large boxes on the left side) or may be degraded. In the case of Notch, the "existent" super-state consists of three concurrent states "domains", "location", and "ligand binding site", indicated by dashed lines. Each of these parallel sub-states has its own (interrelated) dynamics. For example, initially[1] the Notch receptor molecule is located in the "cytoplasm". From this state, a transition may occur either to state "membrane" or to state "nucleus". Both transitions depend on exponentially distributed random waiting times, however, the transition from "cytoplasm" to "membrane" may only fire in case of a mature Notch molecule consisting of both (intracellular as well as extracellular) receptor domains, i.e., only if "domains" is in the "mature" state. Conversely, only the intracellular domain (Nicd) may translocate from the cytoplasm to the nucleus.

A transition from state "membrane" back to "cytoplasm" is triggered by an event "cleavage", which is caused by another state transition describing the cleavage of activated Notch ("bound" state) leading to a loss of its ligand binding site. Please note, events in the original Statecharts formalism are broadcasted. Therefore, the "cleavage" event not only triggers a state change from "membrane" to "cytoplasm", but also from "mature" to "Nicd" and moreover also leads to a change of the receptor binding site of Delta protein from state "bound" to "Necd". Besides facilitating the description of various effects caused by a single event, the asynchronous broadcast communication approach of Statecharts hampers modeling of biochemical reactions like the binding of two proteins, as the sender of an event may change its state although no bind-

---

[1]Initial states are indicated by small black circles.

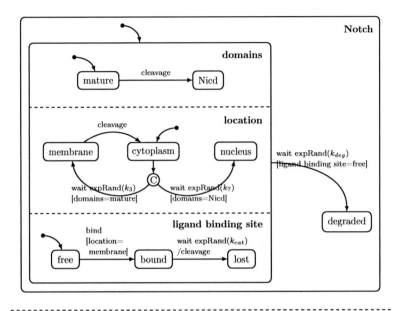

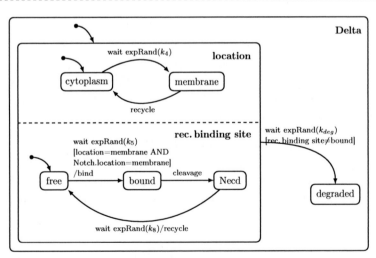

Figure 5.5: Individual-based Statecharts model of reactive Notch and Delta molecules.

ing partner is reacting on it. Also, since numerous receivers may react on the event of one sender, some kind of *resource managing* component may be required that keeps track of the states of each molecule and is responsible for triggering bimolecular reactions (Kam et al. (2001), see also next section on Discrete Event System Specification). In the model of Figure 5.5, by contrast, this is not the case, as in this toy example merely one Notch and one Delta molecule exist which may interact with each other. Binding between both molecules then simply requires to check their "location" states, i.e., the transition from Delta's state "free" to state "bound" may only fire under the condition that both molecules are located within the membrane.

Such an unrealistic example is given here, as the original Statecharts formalism does not support object instantiation and thus would require to describe each individual molecule by a distinct chart. Needless to explain that this approach is highly impractical for modeling biochemical and other systems consisting of large quantities of similar entities. Therefore, object-oriented frameworks like Rhapsody (Harel and Kugler, 2004), GemCell (Amir-Kroll et al., 2008), and Biocharts (Kugler et al., 2010) have been developed enabling an instantiation of multiple objects from classes described as Statecharts. However, this approach requires additional syntactical and semantical means for inter-object communication as well as instantiation and termination of objects (Harel and Gery, 1996), which is typically realized differently by each tool and thereby diminishes the generality and simplicity of according model descriptions.

Alternatively, to remain with the original Statecharts formalism and to avoid the aforementioned artifacts caused by asynchronous broadcast communication, biochemical processes may be described in a more abstract way, in which reactions are represented by concurrent states and pools of distinct molecular species are represented by according global variables (see Figure 5.6). Please notice, initial values of these variables will be set by *entry* actions taking effect when the "cell" state is being entered and dynamic changes will be triggered either by an incoming event message or by the flow of time. Thereby, it is possible to specify arbitrary kinetic rates (see, e.g., the update transition of state R6) and interlevel causation is realized by taking the values of certain

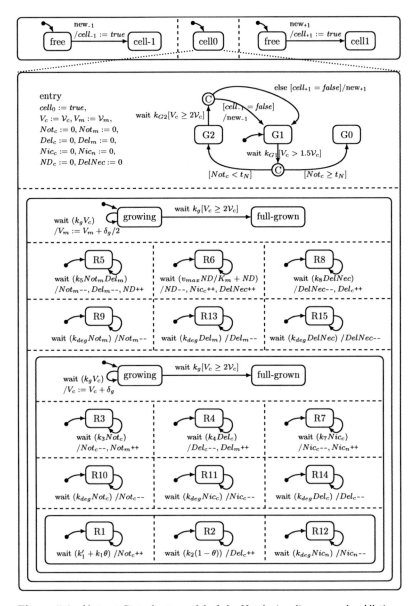

Figure 5.6: Abstract Statecharts model of the Notch signaling example. All time-triggered state changes are assumed to be exponentially distributed waiting times. State names representing biochemical reactions correspond to Table 3.1.

variables into account (e.g., conditional cell cycle phase transitions that are checking compartment volumes and the cytoplasmic Notch amount).

The presented abstract description of biochemical reactions permits to model large quantities of molecular species, however, at the same time we lose the capability of reducing combinatorial complexity, as all variables are globally visible and different conditions of a molecule can no longer be described in terms of concurrent states. Also, since the original Statecharts formalism does not support modeling of variable structures, each potentially appearing cell needs to be described in advance, as it is depicted on top of Figure 5.6. Hierarchical nesting of statecharts is used here to describe the model's structure explicitly, e.g., to make clear that certain reactions are taking place within particular compartments only. However, due to the global scope of generated events in Statecharts, no further functionality is added by the hierarchical structuring, unlike the restricted communication scopes in BioAmbients and cBNGL, for instance. On the other hand, these approaches do not allow to equip structural elements with a state and dynamics of their own, which, by contrast, is supported by Statecharts as there is no distinction between structural and behavioral elements.

5.2.2 Discrete Event System Specification

First introduced by Bernard P. Zeigler (1976), the *Discrete Event System Specification* (DEVS) denotes a classical discrete event formalism with a continuous time base, where events like state changes are taking place at discrete points in time. DEVS shares many similarities with Statecharts (see, e.g., Uhrmacher and Kuttler, 2006; Ewald et al., 2007), such as employing a reactive systems metaphor by focusing on states and state transitions based on incoming events. Rooted in systems theory, the formalism also allows specifying models in a modular and nested fashion yielding a composition hierarchy (Zeigler, 1984, 1987). Therefore, Statecharts or Statechart-like notations are frequently applied to visualize the dynamics of DEVS models (Borland and Vangheluwe, 2003; Risco-Martín et al., 2007).

In DEVS two different kinds of model components exist: *atomic* and *coupled* submodels. Atomic models are the active components, i.e., they describe the model's dynamic behavior. Coupled models, by contrast, serve as passive containers for other coupled as well as atomic model components and define how these components are interconnected with each other, i.e., they define the model's structure. Formally[2], an atomic DEVS model is defined by a tuple

$$M_A := \langle\, X, Y, S, ta, \delta_{\text{ext}}, \delta_{\text{int}}, \delta_{\text{con}}, \lambda \,\rangle$$

where X is the set of inputs, Y is the set of outputs, S is the structured set of states, $ta : S \to \mathbb{R}^{\geq 0} \cup \{\infty\}$ is the time advance function determining the duration of states, $\delta_{\text{ext}} : Q \times X \to S$ is the external transition function determining how inputs change the state of the model, with state set $Q = \{(s, e)| s \in S, 0 \leq e < ta(s)\}$ including the time interval that has been elapsed since the last event, $\delta_{\text{int}} : S \to S$ is the internal transition function determining how the state of the model changes when the specific time interval associated with state $s \in S$, i.e., $ta(s)$, has been elapsed, $\delta_{\text{con}} : S \times X \to S$ is the confluent transition function determining state changes for situations in which internal and external events coincide, and $\lambda : S \to Y$ is the output function determining which outputs are generated when an internal state transition occurs.

So, similar to Statecharts, state transitions in DEVS can be either triggered by external events or by the flow of time and likewise both deterministic as well as stochastic time intervals with arbitrary kinetics are possible. However, unlike Statecharts, output events may be generated only in case of an internal – i.e., time-triggered – state transition, leading to artifacts if a model requires to describe an instantaneous response to an incoming message. The interactions themselves are timeless and they are constrained at the superior organizational level, i.e., which messages may reach which component is specified by coupling sets being part of the enclosing coupled model component. Therefore, a coupled DEVS model is formally defined by a tuple

$$M_C := \langle\, X, Y, D, M_i, EIC, EOC, IC \,\rangle$$

where X and Y are again sets of inputs and outputs respectively, D is the name set of enclosed components, M_i with $i \in D$ is the structured set of sub-components consisting of atomic and/or coupled DEVS models, $EIC \subseteq X \times X_i$

---

[2]The P-DEVS variant is presented here, which includes minor additional features compared to the original DEVS formalism (see Chow and Zeigler, 1994; Zeigler et al., 2000).

with $i \in D$ is the set of external input couplings, $EOC \subseteq Y_i \times Y$ with $i \in D$ is the set of external output couplings, and $IC \subseteq X_i \times Y_i$ with $i \in D$ is the set of internal couplings, i.e., couplings that do not connect to the coupled model's environment but define communication links between sub-components M_i only. Please note, in general the input and output sets of DEVS models are plain sets of messages. In practice, however, X and Y are typically structured into typed ports (Zeigler et al., 2000), e.g., X may formally be structured into a set of pairs $\{(i, v) \,|\, i \in InputPorts, v \in X_i\}$, where $InputPorts$ is the set of input ports and each X_i is the set of allowed values for all $i \in InputPorts$ (see example in Figure 5.7).

Output messages in DEVS are sent to all model components connected to the respective output port, which resembles the broadcast communication of Statecharts. Likewise, the original DEVS formalism also does not support variable structures, so that the initial model composition and couplings between submodels are fixed. Therefore, similar to Statecharts, modeling biochemical processes with DEVS typically implies to describe the system from a macro perspective, where the molecular species amounts are represented by distinct state variables and their dynamic changes are described by an ad hoc implementation of a numerical integration or Gillespie's stochastic simulation algorithm, like it is indicated by Figure 5.8 (see also Uhrmacher et al., 2005; Uhrmacher and Priami, 2005; Uhrmacher and Kuttler, 2006; Ewald et al., 2007; Maus et al., 2008). Thereby, however, we might easily run into problems due to combinatorial complexity and one of the main strengths of DEVS – namely to describe a system in a modular and hierarchically structured manner – seems to be reserved for describing dynamic processes in which only few entities are involved, such as gene-regulatory processes or processes at higher organizational levels like entire cells. However, even there the missing support for variable structures denotes a major limitation for modeling biological systems, as, for example, each cell of a multicellular system and each potential interaction between different cells need to be explicitly described in advance and thus modeling of phenomena like cell proliferation is hampered.

To overcome such restrictions, diverse variants of the DEVS formalism have been developed that support the description of variable structure models, e.g.,

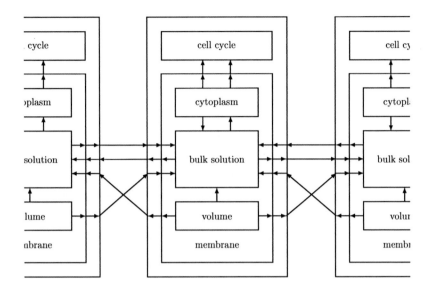

$$
\begin{aligned}
membrane \quad &:= \quad \langle\, X, Y, D, EIC, EOC, IC \,\rangle \\[4pt]
X \quad &:= \quad \{\, (?LVol_m, \mathbb{R}), (?RVol_m, \mathbb{R}), (?LNot_m, \mathbb{N} \cup \{\text{`bind'}, \text{`unbind'}\}), \\
& \qquad (?RNot_m, \mathbb{N} \cup \{\text{`bind'}, \text{`unbind'}\}) \,\} \\[4pt]
Y \quad &:= \quad \{\, (!LVol_m, \mathbb{R}), (!RVol_m, \mathbb{R}), (!LNot_m, \mathbb{N} \cup \{\text{`bind'}, \text{`unbind'}\}), \\
& \qquad (!RNot_m, \mathbb{N} \cup \{\text{`bind'}, \text{`unbind'}\}), (!Vol_c, \mathbb{R}), (!Not_c, \mathbb{N}) \,\} \\[4pt]
D \quad &:= \quad \{\, cytoplasm, bulksol, volume \,\} \\[4pt]
EIC \quad &:= \quad \{\, (self.?LVol_m, bulksol.?LVol_m), (self.?RVol_m, bulksol.?RVol_m), \\
& \qquad (self.?LNot_m, bulksol.?LNot_m), (self.?RNot_m, bulksol.?RNot_m) \,\} \\[4pt]
EOC \quad &:= \quad \{\, (volume.!LVol_m, self.!LVol_m), (volume.!RVol_m, self.!RVol_m), \\
& \qquad (bulksol.!LNot_m, self.!LNot_m), (bulksol.!RNot_m, self.!RNot_m), \\
& \qquad (cytoplasm.!Vol_c, self.!Vol_c), (cytoplasm.!Not_c, self.!Not_c) \,\} \\[4pt]
IC \quad &:= \quad \{\, (volume.!Vol_m, bulksol.?Vol_m), \\
& \qquad (bulksol.!ToCyt, cytoplasm.?FromMem), \\
& \qquad (cytoplasm.!ToMem, bulksol.?FromCyt) \,\}
\end{aligned}
$$

Figure 5.7: DEVS model of the Notch signaling example. (Top) Schematic of model composition and couplings. Lower-level components enclosed by "cytoplasm" are not shown. (Bottom) Specification of the coupled "membrane" model. Please note, input and output ports are indicated by prefixes "?" and "!" respectively.

$bulksol$:= $\langle X, Y, S, ta, \delta_{\text{ext}}, \delta_{\text{int}}, \delta_{\text{con}}, \lambda \rangle$

X := $\{ (?LVol_m, \mathbb{R}), (?RVol_m, \mathbb{R}), (?Vol_m, \mathbb{R}), (?FromCyt, \{\text{'Notch'}, \text{'Delta'}\}),$
$(?LNot_m, \mathbb{N} \cup \{\text{'bind'}, \text{'unbind'}\}), (?RNot_m, \mathbb{N} \cup \{\text{'bind'}, \text{'unbind'}\}) \}$

Y := $\{ (!LNot_m, \mathbb{N} \cup \{\text{'bind'}, \text{'unbind'}\}), (!RNot_m, \mathbb{N} \cup \{\text{'bind'}, \text{'unbind'}\}),$
$(!ToCyt, \{\text{'Delta'}, \text{'Nicd'}\}) \}$

S := $\{ (Vol, VolL, VolR, Del, Not, NotL, NotR, NotDelL, NotDelR, DelNec),$
$NextReaction, TimeToNextReaction \mid$
$Vol, VolL, VolR \in \mathbb{R};$
$Del, Not, NotL, NotR, NotDelL, NotDelR, DelNec \in \mathbb{N};$
$NextReaction \in \{DegNot, DegDel, BindingL, BindingR,$
$CleavageL, CleavageR, DegDelNec, RecycleDel\};$
$TimeToNextReaction \in \mathbb{R}^{\geq 0} \cup \{\infty\} \}$

$ta(s, NextReaction, TimeToNextReaction)$:= $TimeToNextReaction$

$\delta_{\text{ext}}((s, NextReaction, TimeToNextReaction), e, x)$:=
    switch(x)
        case $?LVol_m$: $newstate := s$ with $VolL :=$ getValue($?LVol_m$);
        case $?FromCyt$: if getValue($?FromCyt$) = 'Notch'
                then $newstate := s$ with Not++
                else $newstate := s$ with Del++;
        . . .
    end switch
    ($newstate$, nextReact($newstate$), nextReactTime($newstate$))

$\delta_{\text{int}}(s, NextReaction, TimeToNextReaction)$:=
    switch($NextReaction$)
        case $DegNot$: $newstate := s$ with Not--;
        case $BindingL$: $newstate := s$ with Not--, Del--, $NotDelL$++;
        . . .
    end switch

$\lambda(s, NextReaction, TimeToNextReaction)$:=
    switch($NextReaction$)
        case $DegNot$: ($!LNot_m, Not - 1$),($!RNot_m, Not - 1$);
        case $BindingL$: ($!LNot_m$, 'bind');
        . . .
    end switch

Figure 5.8: Atomic DEVS model of the bulk solution component. Please note, the confluent transition function δ_{con} is omitted, as it describes nothing else than the simple successive execution of δ_{int} and δ_{ext}. The partial state set s is defined as $s := S \backslash \{NextReaction, TimeToNextReaction\}$.

dynDEVS (Uhrmacher, 2001) and ρ-DEVS (Uhrmacher et al., 2006). Model components in dynDEVS represent a set of models that may generate themselves by according transitions. Each element of the set thereby represents an individual instance of the model, which may also dynamically change its individual couplings to other components, i.e., its interaction capability. ρ-DEVS goes even further and extends dynDEVS by variable interfaces (port sets) and multi-couplings, by which the communication structures between components are automatically changed depending on the availability of ports. The idea is to let variable interfaces signalize significant state changes to a model's environment and thereby change its interaction capabilities, which can be straightforwardly used, e.g., to model protein-protein interactions that become only possible due to certain intramolecular modifications like the phosphorylation of specific amino acid residues.

While variable structure approaches generally facilitate the description of models at a detailed (micro) level of abstraction, their modular composition capacity similar to that of regular DEVS can be also used for combining micro and macro perspectives within the same model, as has been done by Degenring et al. (2003, 2004), for instance. However, none of the aforementioned approaches eliminates another major limitation for multilevel modeling with DEVS, namely that coupled models do not have a state and behavior of their own. Unlike hierarchical Statecharts but similar to ambients in BioAmbients and compartments in cBNGL, coupled DEVS models are simple containers and their behavior is thus completely determined by their enclosed entities and the way these entities are coupled with each other. In case of describing processes that depend on the overall state of the system, this passive fate of coupled models contradicts an intuitive structural organization of the model, as high-level properties need to be described by atomic model components residing at the same level like those components that are influenced by them via downward causation. Hence, the model's composition hierarchy and the natural perception of the system disagree. Also, downward and upward causation between the respective components have to be realized by sending events asynchronously, which is an inappropriate approach for taking side-effects into account and burdens the modeling significantly. Therefore, aiming at explicit

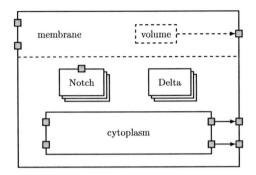

Figure 5.9: Abstract structure of a coupled ML-DEVS model describing the cellular membrane compartment. Unlike other DEVS variants, ML-DEVS allows describing high-level properties like the dynamic compartment volume as part of the coupled model (dashed elements). Sub-components like Notch and Delta molecules signalize crucial state changes to the coupled model via adding or removing certain ports (small grayish squares).

multilevel modeling in a DEVS regime, another extension has been developed called ML-DEVS.

With ML-DEVS (Uhrmacher et al., 2007; Steiniger et al., 2012), coupled models are equipped with an own state and behavior, so that high-level properties of a system can be described at the right place. For instance, the volume of a compartment and its dynamic change can now be part of the coupled model description (cf. Figure 5.9) rather than described by a separate sub-component. ML-DEVS also introduces explicit means for interlevel causation via information propagation and activating events[3]. Inherited from ρ-DEVS, model components in ML-DEVS may change their ports and thereby signalize crucial state changes to the outside world. Upward causation is supported, as the coupled model has an overview of the number of enclosed components being in a particular state – i.e., exhibiting a particular set of ports – and to take this into account when updating its state. An invocation of the coupled model's state transition function can be triggered by *activation constraints*, e.g., if the number of components being in a certain state surpasses a certain threshold.

---

[3]See Uhrmacher et al. (2007) and Maus et al. (2008, Sec. 3.4) for detailed formal definitions and concrete examples.

In the downward direction, causation can be realized by accessing global state variables via *value couplings*. Therefore, global information is mapped to specific port names and each enclosed model component may directly access the according variables by defining input ports with corresponding names. In addition, the coupled model may also activate its sub-components by sending them events. Thereby, it becomes possible to let several components interact synchronously, which is a more natural way of describing biochemical reactions compared to the conventional asynchronous communication. However, like other DEVS variants, ML-DEVS model encodings still tend to be large and difficult to comprehend, which clearly has a negative effect on the formalism's accessibility. Therefore, although highly expressive, no DEVS variant seems overly suited for modeling ordinary biochemical reactions. Only if entities show rich internal dynamics, the reactive systems view of DEVS appears fitting (Maus et al., 2008).

5.2.3 Bigraphical Reactive Systems

A *bigraph* combines two different kinds of graphs – a set of *trees* (i.e., a forest) and a *hypergraph* – to form a mathematical structure that represents both nesting as well as linkage of nodes at the same time (Figure 5.10). Both subgraphs of a bigraph therefore consist of the same set of nodes, however, these nodes may be linked differently to represent different relationships among the nodes, such as locality and interaction. Developed by Robin Milner (2009), a comprehensive mathematical theory behind bigraphs allows to study mobile and concurrently communicating systems within a well-defined formal framework.

Each node of a bigraph has assigned a *control* determining the number of *ports* to which links can be attached. Bigraphs can be built in a modular manner by composition of several smaller graphs. Interfaces thereby determine how and which distinct bigraphs can be composed. The *outer face* of a bigraph is defined by a set of outer names and a set of root nodes (*regions*, while the *inner face* is defined by a set of inner names and a set of *sites*. Two bigraphs may be composed if the outer face of one graph equals the inner face of the other, i.e., inner and outer names as well as regions and sites must coincide. Composition can be used to specify graph rewriting rules for modeling the

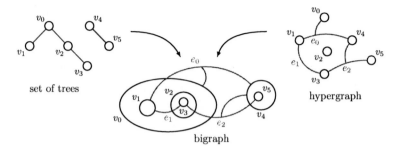

set of trees

bigraph

hypergraph

Figure 5.10: The bigraphical structure of bigraphs.[4]

dynamic change of a bigraph in terms of rule *schemata*, so that only small parts applicable to various concrete structures need to be described. Although strongly different from reactive systems approaches like Statecharts and DEVS, this graphical rule-based modeling approach is commonly termed *Bigraphical Reactive System* or BRS (Milner, 2001, 2002, 2006). A stochastic semantics has been defined to study the evolution of BRS models quantitatively (Krivine et al., 2008).

Let us consider an example. To model the compartmentalized biochemical Notch signaling processes presented in Chapter 3, first a set of controls

$$\mathcal{K} = \{\, \text{Cell}:3, \text{Nucleus}:0, \text{Nicd}:1, \text{Necd}:2, \text{Delta}:2 \,\}$$

is specified defining names and linkage arities of the model's entities, i.e., the graph nodes. Made up of controls "Cell" and "Nucleus", the initial bigraph

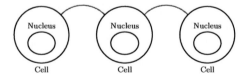

serves as a starting point on which the set of graph rewriting rules shown in Figure 5.11 can be applied.

An application of a certain rule requires instantiation by composing the left-hand side bigraph of the rule (redex) with *context* and *parameter* graphs

---

[4]Modified illustration from Milner (2009).

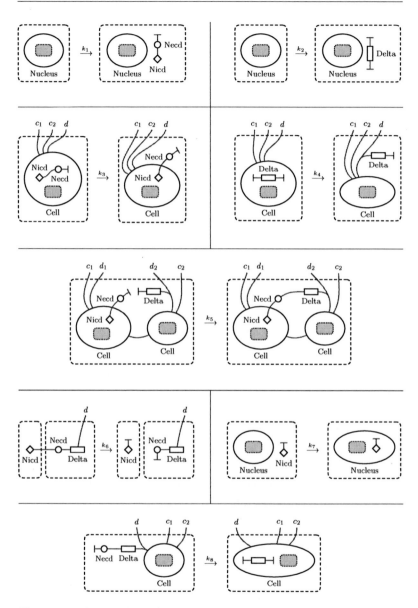

Figure 5.11: Reactive Bigraphs model of the biochemical Notch signaling processes. The presented rules correspond to reactions 1–8 in Table 3.1. Protein degradation reactions are omitted.

115

that produce exactly the current state of the model. Regions are thereby represented by dashed rectangles around defined controls and sites are represented by small grayish rectangles. For example, the redex of the first rule of Figure 5.11 inserted into a context graph

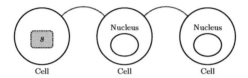

with a single site s would match the initial state and may thus be applied. The result would be a produced Notch molecule within the first cell represented by linked nodes "Nicd" and "Necd", as shown on the right-hand side of the first rule. Please note, the rule may be instantiated in three different ways, since composition with three different context graphs would produce the initial graph consisting of three cells. The remaining rules of Figure 5.11 – some of which comprising link names on top outside of regions expanding the interface by outer names – are instantiated similarly. Unlike Statecharts and DEVS, Bigraphical Reactive Systems thus pursue an alternative hierarchical modeling approach, where the dynamics, i.e., the reaction rules, are specified more or less independently from the concrete model structure similar to conventional flat rule-based approaches like BNGL.

Due to its inherent support for nesting and visual graph rewriting rules, a BRS allows for intuitively modeling hierarchically organized systems with dynamic structures like the movement and merging of complex sub-graphs. Thereby, each hierarchical level may has its own state and its own behavior, however, unlike entities in attributed languages like colored Petri nets and BNGL, the states of nodes in a bigraph are determined by their linkage to other nodes only. Arbitrary (numerical) attributes are not supported. Consequently, describing high-level dynamics like changing compartment volumes and thus also the description of interlevel causation is hampered. Moreover, the stochastic Bigraphs formalism relies on the law of mass action (Krivine et al., 2008) and does not support to constrain reaction rates flexibly, which additionally limits the formalism's applicability for systems biology in general and multilevel modeling in particular.

5.3 Concluding Remarks

An overview of diverse hierarchical modeling approaches has been given in this chapter. Thereby, the Notch signaling model introduced in Chapter 3 served again as a running example to discuss their support or suitability for describing diverse aspects of biological multilevel systems. A brief synopsis and comparison between the different approaches is given in Table 5.2.

All modeling languages presented in this chapter allow for structuring models in a hierarchically nested manner. However, different strategies are employed to achieve this feature. In the case of hierarchical Petri nets, BioAmbients, and cBNGL, existing flat languages have been extended by means of hierarchical structures. This strategy seems to be reasonable, as these languages have already proven their value for describing diverse flat models. On the other hand, the newly introduced structural elements often appear to be rather synthetic constructs with limited functionality. Other hierarchical modeling approaches have hierarchical structure included from the onset. Consequently, the support for describing hierarchically structured models is an inherent property of these languages and thus they typically put the focus on nesting and composition.

Whether a particular language supports hierarchies inherently or whether it is an extension of a pre-existing flat approach says nothing about its capability to describe dynamic model structures. Both BioAmbients and BRS allow for describing dynamic structures by creating, deleting, moving, and merging structural model entities. By contrast, neither the original Statecharts formalism nor DEVS and hierarchical Petri nets support dynamic structure modeling (although this may be mimicked to some degree by complex events or colored tokens, for instance). Compartments in cBNGL are also fixed, however, inherited from the original BioNetGen language, the approach still allows for modeling dynamically changing molecular bonds.

An appropriate representation of states and behavior at different levels is another aspect that must not necessarily be enabled by supporting hierar-

Table 5.2: Summary of the expressivity of diverse hierarchical modeling approaches with respect to describing certain multilevel aspects. Good and moderate support is denoted by "+" and "o" respectively, while "−" denotes no or only poor support for modeling the according aspects.

	Hierarchical model structure	Dynamic model structure	State & behavior at diff. levels	Arbitrary rate laws & constraints	Upward & downward causation	Space beyond compartments
Hierarchical colored Petri nets	+	−	o	+	o	+
BioAmbients	+	+	−	−	−	o
Compartmental BNGL	+	o	o	+	o	−
Statecharts	+	−	+	+	+	o
Discrete Event System Specif.	+	−	o	+	o	+
Stochastic Bigraphs (BRS)	+	+	−	−	−	+

chies. While the majority of languages presented here allow for assigning arbitrary attributes to a model's components and thus for representing states quite differently depending on the dynamic processes to be described, only with Statecharts and Bigraphs it is possible to equip the containing structural elements with a state and behavior of their own. The latter formalism, however, puts the focus on the linkage between components (other attributes are not supported), which hampers multilevel modeling seriously, as, e.g., describing dynamic compartment volumes is impossible.

Similarly serious limitations for multilevel modeling arise by the missing support for arbitrary rate kinetics and reaction constraints in BioAmbients and BRS. As a consequence – and due to the limited capabilities for representing different states – these two formalisms also support interlevel causation only poorly. In Statecharts, by contrast, side-conditions can be easily taken into account due to constraining transitions flexibly and by accessing diverse states from other organizational levels via broadcast messaging and globally visible variables. Thereby, however, other aspects become affected, like the succinctness of model descriptions.

Spatial structures beyond nested compartmentalization, i.e., the discrete neighborhood of different cells in our example, are well-supported by either constraining reactions based on attributes and according mathematical expressions (colored Petri nets) or by defining the neighborhood explicitly by connecting the respective components via links (DEVS and Bigraphs).

Apart from the discussed multilevel aspects, the examples in this chapter also indicate diverse suitability of the different approaches for modeling dynamic processes at different levels. A reactive systems metaphor – as pursued by Statecharts and DEVS – is well suited for describing dynamics at cellular and cell population levels, such as the proliferation of a cell influenced by internal and external events. However, due to its asynchronous communication pattern and the lack of an inherent macro view on the model's component states for determining correct reaction rates, the approach requires considerable extra efforts for modeling biochemical reactions. Hence, a reaction-centric modeling approach, such as Petri nets, BNGL, or BRS, seems to be the better choice for describing such processes.

Chapter 6

A Language Concept for Accessible Multilevel Modeling

In the previous two chapters we have seen different modeling approaches and how they can be applied in order to describe biological systems at multiple levels. Thereby, each approach has its advantages and disadvantages with respect to modeling biological systems and different organizational levels in general and multilevel behavior in particular.

In this chapter, the concept of a novel modeling approach is presented, which aims at facilitating multilevel modeling in systems biology. Therefore, first a set of useful requirements is identified, which is based to a large extent on the previously discussed aspects and strategies for modeling biological multilevel systems. Subsequently, in Chapter 6.2, a tailor-made multilevel modeling concept is introduced step by step and finally a concrete realization of this concept is presented.

6.1 Requirements

6.1.1 General Modeling Paradigm

The design of a domain specific modeling language should be strongly influenced by its intended application, i.e., the chosen modeling paradigm should map on the kinds of systems to be modeled in a straightforward and suitable manner. Thereby, among other criteria, finding the right balance between straightforwardness and expressive power is crucial for achieving a satisfying accessibility degree. In case of designing a multilevel modeling language, this may be a challenging task due to the possibly highly diverse nature of different organizational levels, ranging from low-level *molecular dynamics*, over *biochemical reaction networks, cells*, and *tissue dynamics*, up to *whole organism's physiology* in the application field of systems biology.

However, multilevel models typically do not consider the entire hierarchy of an organism but focus on just a few organizational levels. Thereby, a *middle-out* strategy has proven to be a pragmatic approach (Kohl et al., 2010). Denis Noble explains the approach as follows: *"Biological function happens at different levels. We can gather quantitative data at any level. Once we have enough of it to feed into a simulation, we can start a systems analysis at that level. [...] Then, when we have established sufficient understanding and success at our chosen level, we can reach out [...] to other levels."* (Noble, 2006, p. 79f). Hence, the decision for a specific paradigm or modeling metaphor should clearly depend on the most prevalent levels of interest, assuming that they typically denote the starting points of multilevel modeling projects. However, since multilevel modeling can start at any level, the question is which are the most important levels of interest?

Taking the number of published articles and models in publicly available repositories – like the BioModels Database – as a measure for the relevance of certain levels in systems biology, some levels appear to be more often in the focus of investigation than others. In particular the level of *biochemical interactions* seems to be the most important level under study, but also processes at *cellular* and *cell population* levels are frequently found subjects of investigation. Hence, a general purpose modeling approach for describing biological multilevel models should particularly aim for handily expressing dynamic processes at these intermediate levels of organization. The examples from the previous chapters indicate that two different general approaches are notably qualified to map on these levels of interest and therefore come into consideration: the object-centered *reactive systems* metaphor (e.g., DEVS and Statecharts) and a *reaction-centric* perception of a model (e.g., Petri nets and BNGL).

As Fisher et al. (2011, p. 72) point out, *"a living cell [...] is not only reactive in nature, but is the ultimate example of a reactive system, and so are collections thereof"*. Reactive systems are thus well qualified to represent processes at the cellular and cell population levels, such as the traversal through the cell cycle or communication between individual cells. On the other hand, we have learned that it is less straightforward to describe biochemical processes by using the reactive systems metaphor (see, e.g., Ewald et al. (2007) and Chapter 5, pages 102–113 of this thesis).

A suitable and straightforward approach for the formal description of biochemical dynamics in turn is to apply a reaction-centric modeling paradigm (Vass et al., 2006; Danos, 2009; Heiner and Gilbert, 2011). Unlike the π-calculus and other process algebras, which have also demonstrated their suitability for modeling biochemical reactions (Priami et al., 2001; Calder et al., 2006a; Cardelli, 2007; John, 2010), here the objects of interest are entire reactions rather than individually interacting processes. Compared to the process interaction view, the reaction-centric paradigm is not only closer to the textbook notation of chemical reactions and thus probably more intuitive, in Chapter 4 it has also been shown that the approach facilitates an abstraction from elementary reactions and the description of interlevel causation by taking certain side effects into account.

Beyond biochemistry, the reactions metaphor suits also various dynamics at other organizational levels. In a 2005 bulletin board discussion[1] on encoding non-biochemical models in the Systems Biology Markup Language (SBML; Hucka et al., 2003) – which is a widely-used reaction-centric model exchange format – Nicolas Le Novère put it in a nutshell: *"As far as the same rate-law can describe the behaviour of a population of molecules, a population of cells, or a population of individuals, I don't see the problem. Lotka-Volterra is a good example."* Moreover, as also shown by the activity scanning approach (Hooper, 1986), reactions may serve as a metaphor for state transition systems in general. Figure 4.9 on page 85, for example, shows how to describe discrete cell cycle phase transitions by using the reaction-centric BioNetGen Language.

Hence, a reaction-centric modeling paradigm seems to qualify as a suitable starting point for accessible multilevel modeling in systems biology, as it is

---

[1] http://sbml.org/Forums/index.php?t=tree&goto=1757 (accessed Sept. 25th, 2012)

well-suited for describing biochemical reactions and in addition also permits to appropriately describe certain dynamic state changes at other organizational levels. However, it remains to be investigated whether this metaphor applies well to processes at every level of organization, since, e.g., *"...many cell biological phenomena require calculation of structural dynamics, deformation of elastic bodies, spring-mass models and other physical processes"* (Kitano, 2002a, p. 209) that might be difficult to describe by applying the reactions metaphor. Therefore, an integration of such complex physico-mechanical dynamics into the operational semantics of a language might be required to model respective phenomena (see, e.g., Michel et al., 2009; Bittig et al., 2011).

6.1.2 Model Structure

As has been shown in Chapter 4, a lack of structuring elements allows describing hierarchical relationships only implicitly, making it difficult to capture the essential characteristics of a multilevel model without further explanation. According to Bruce Edmonds (1999, p. 34), *"a good modelling language will not only be expressive enough to clearly specify the required possibility spaces, but also it will have explicitly defined relations that systematically reflect the corresponding relations between the possibility spaces"*. Accessible multilevel modeling therefore requires to make the hierarchical structure of biological systems explicit.

To appropriately reflect our natural perception of composition hierarchies, the modeling language should support nested model structures, i.e., parts nested within wholes, allowing for both a vertical as well as horizontal separation of entities. Moreover, a support for modeling dynamic structures is strongly desired, as the hierarchical composition of biological systems is changing frequently due to phenomena like fusion of compartmental membranes, molecular translocation and cellular migration, or cell division and death, for example.

Since dynamic behavior can be found at any level, it is also important to allow for modeling entities at any organizational level that have a state and behavior on their own. That means, unlike many formalisms supporting hierarchical model structures, e.g., DEVS and cBNGL, in which upper hierarchies are simple containers, dynamic multilevel behavior must not be restricted to atomic entities, i.e., entities that denote the leaf nodes of a tree graph that is representing the model's hierarchy and therefore do not enclose further components. The multilevel modeling language should rather also allow to equip coupled entities consisting of a set of other entities with an own state that may be dynamically changed according to certain rules.

These states furthermore need to be accessible from other levels in order to describe interlevel causation. Thereby, context-dependent availability should be favored over globally visible states, as otherwise according variables need to be defined from the beginning – in order to avoid name collisions – and this may significantly decrease the succinctness (see also Section 6.1.4) of model descriptions and hampers modeling of structural changes (cf. Chapter 4.2).

6.1.3 Other Expressiveness Requirements

Besides the need for explicitly representing dynamically changing nested hierarchies, modeling complex multilevel systems in a straightforward manner imposes other special requirements in respect of the expressive power of the language the model shall be described with.

Different representations of states

Informally, the expressive power (or the expressiveness) of a particular language can be seen as the theoretical or practically feasible ability to express a certain set of ideas or facts in that language (Edmonds, 1999). In the case of multilevel modeling, due to the diverse nature of observed data and the need to make different abstractions at different organizational levels, the applied language should allow for expressing states or conditions also quite differently. For example, the amounts of molecular species may be described best by discrete integer values, while other state variables, e.g., the volume or the pH of a cellular compartment, may be represented more appropriately by real numbers. Also representing certain states by Boolean values or arbitrary character strings may be useful for an appropriate description and to enhance the accessibility.

Arbitrarily constrained transitions

Pertaining to dynamic state changes, the multilevel modeling language needs to be sufficiently expressive to allow for making various (behavioral) abstractions – like combining multiple elementary reactions in one step – and in particular for describing interlevel causation. Therefore, supporting dynamic processes with arbitrary rate kinetics is required, but also constraining processes flexibly and based on diverse state variables – like it is supported, e.g., by the attributed π-calculus and colored Petri nets – is an essential feature for modeling multilevel systems. Thereby, states at the same level of organization may need to be taken into account, just like states at lower or higher levels. Also, for taking such side effects into account, it is often necessary to define n-ary reactions comprising of more than two reactants, i.e., indispensable entities taking part in the dynamic process. Hence, the language must not be restricted to binary interactions, like it is the case in the π-calculus, for instance.

Spatial dynamics

Aside from nested model structures, different kinds of state representations, arbitrary rate kinetics, and flexible reaction constraints, the language should also – at least rudimentary – support the representation of space, since spatial aspects beyond simple compartmentalization play an important role in many multilevel modeling studies.

Howsoever, despite the clear requirement for expressing the above multilevel aspects, all in all we need to also carefully consider the inevitable trade-off between the expressive power of a language and its ability to be analyzed. The computational complexity for simulating a model typically increases with the formalism's expressiveness. Therefore, simulation in a reasonable amount of time may become impractical if the language is "too" expressive.

6.1.4 Succinctness and Compositionality

Other language requirements stem from the tendency of multilevel models to become rather large and difficult to comprehend. That means, by applying conventional approaches, many lines of code or a large number of reactions as well as intricate definitions of dynamic processes may be needed in order to describe the according multilevel behavior. Therefore, to keep multilevel models small and manageable, the language needs to be succinct and should allow for extending a model incrementally by composing smaller parts.

The succinctness or conciseness of a language determines how compact a description of a certain set of ideas or facts in that language might become. Hence, it strongly influences the psychological complexity and thereby its accessibility. A formal analysis of the succinctness is typically done in comparison to other languages, i.e., succinctness is a relative measure (Hartmanis, 1979) in dependence on the expressiveness: *"Of two equally expressive languages, one may be more succinct than the other."* (Henzinger et al., 2009, p. 14). Studying the expressivity of diverse languages has also led to a "conciseness conjecture", stating that *"programs in more expressive programming languages that use the additional facilities in a sensible manner contain fewer programming patterns than equivalent programs in less expressive languages."* (Felleisen, 1991, p. 70).

In general, an effective approach for achieving succinct and concise model descriptions is to avoid redundancy, e.g., by instantiating concrete objects from a generalized class. In the realm of reaction-centric modeling languages, redundancy may be effectively reduced by applying reaction patterns or schemata. For example, applying a rule-based approach like the κ-calculus or BNGL may yield rather compact and concise models compared to classical (non-schematic) reaction networks, as has been demonstrated by Danos (2009), for instance.

Another approach that helps keeping large models to be manageable is composition. A modeling language that supports a *"notion of compositionality [...] allows the user to incrementally define the whole set of systems of interest by adding information to the part already developed."* (Uhrmacher and Priami, 2005, p. 323). Compositional languages like the π-calculus, BNGL, and DEVS thus allow for a *divide and conquer* strategy, by which a system may be described in terms of small modules that can be combined afterwards. Small parts are considered to be more easy to understand and also to maintain than large ones. Hence, composition is considered to facilitate modeling of highly complex systems and therefore the multilevel language should also support some notion of compositionality.

6.2 Concept

As has been discussed in detail above, the concept for an accessible multilevel modeling language presented in this chapter needs to fulfill or implement the following requirements:

- Reaction-centric modeling paradigm

- Dynamic nested model structure with states and behavior at any level

- Diverse representations of states and conditions

- N-ary reactions with arbitrary rate kinetics and flexible constraints

- Simple spatial dynamics

- Succinctness

- Compositionality

These requirements have led to the development of ML-Rules (Maus et al., 2011) – a rule-based approach for quantitative modeling of biological systems at different integrated levels of an organizational hierarchy. The language concept is presented in terms of an informal description of the syntax and semantics with the help of several simple examples. At first some general properties of the language will be discussed before it comes to presenting more detailed aspects. Thereby, we start with flat examples and aspects that are not directly related to hierarchical systems modeling to successively approach concepts that are obviously related to describing multilevelness, i.e., hierarchical nesting, states and behavior at any level, as well as upward and downward causation between different levels.

6.2.1 General Semantics

The general semantics of ML-Rules, i.e., its underlying formal mathematical model, is based on continuous-time Markov chains (CTMCs). That means, like in several of the previously mentioned approaches, e.g., stochastic Petri nets and different variants of the stochastic π-calculus, we have a possibly infinite number of states of our modeled system, where each transition from one state to another follows an exponentially distributed random waiting time (Wilkinson, 2006). Model states are represented by multisets of species (distinguishable model entities), i.e., the semantics of states in ML-Rules is discrete population-based, which means that the amounts of species are represented by natural copy numbers rather than real-valued concentrations.

Describing biochemical reaction networks stochastically in terms of CTMCs is a widely applied methodology in systems biology (Wilkinson, 2006; Ullah and Wolkenhauer, 2011). The reason for choosing a stochastic semantics for our multilevel modeling language lies in the observation that at higher levels of organization (like cells) no longer abundant numbers will be able to balance fluctuations, as can be often observed at lower organizational levels, for instance, proteins and other molecules that are involved in metabolic pathways. But also in the case of intercellular as well as intracellular signal transduction pathways (Bhalla, 2004; Calder et al., 2006b) or when describing dynamic processes at levels further down in the hierarchy, e.g., gene regulatory processes (de Jong, 2002), stochastic events may play crucial roles due to low copy numbers of involved species. Hence, *"it should be noted that stochastic effects are not only important on the molecular scale"* (Meier-Schellersheim et al., 2009, p. 8) and therefore stochasticity is often an essential feature for multilevel modeling of biological systems (Brook and Waters, 2008; Lavelle et al., 2008; Twycross et al., 2010; MacNamara and Burrage, 2011). However, it should be also noted that stochasticity is not necessarily constrained to CTMCs, as sometimes at higher levels other than exponential time delays are required, e.g., normal distributions in the case of modeling cell cycle dynamics (Walker et al., 2004). For such settings, a viable solution might be the definition of an alternative operational semantics supporting generally distributed firing rates, as has been proposed by Mura et al. (2009).

6.2.2   Rule-based Modeling Approach

ML-Rules is not the first rule-based modeling approach designed for systems biology. Therefore, instead of completely reinventing the wheel, the language employs proven concepts and thus shares several similarities with existing solutions, in particular with React(C) (John et al., 2011). However, before we go into detail by presenting the modeling concept and thereby also discussing related approaches, here at first the general decision for selecting a rule-based modeling strategy will be motivated.

Diverse reasons are underlying this decision. Firstly, rule-based approaches employ a *reaction-centric* modeling paradigm, which complies with our first requirement (see page 121ff) and allows for a notation along the lines of well-known chemical reaction equations, making the language generally accessible to domain experts in the field of systems biology. Moreover, the effectiveness of modeling complex biological systems with rule-based approaches has been repeatedly demonstrated, in particular in cases of modeling signal transduction pathways (Hlavacek et al., 2006) and metabolic reaction networks (Cohen and Bergman, 1994; Faeder et al., 2009) at the molecular level. This effectiveness primarily relies on the *succinctness* of models, achieved by *attributed entities* and the general conceptual modeling and simulation strategy – or "world view" (Hooper, 1986) – into which rule-based approaches can be categorized, namely *activity scanning*.

An activity scanning approach comprises a set of independent "waiting" modules and uses condition testing to decide which of the modules may be executed next (Balci, 1988). In a rule-based approach, these modules are represented by *rules*, each of which consists of a condition that needs to be fulfilled to perform a certain action, e.g., the creation, modification, or removal of certain model entities. According instructions are also part of the rule. In most rule-based languages, entities may be equipped with different attributes or states. Thereby, conditions may be specified in terms of *conditional patterns*, so that a single rule may encode for various contexts and thus becomes a rule *schema*. This way, rule-based modeling allows for describing models rather succinctly. In addition, due to the modular approach with rules as independent components, the compositionality requirement is inherently enabled and also modeling of variable structures is facilitated, as the approach permits to dynamically reconfigure a model's composition.

Furthermore, although not exclusively found in rule-based settings, (colored Petri nets and the attributed π-calculus, for instance, are examples from other classes of modeling languages), a rule-based language design permits to flexibly constrain reactions and thereby to increase its expressivity, as has been shown by John et al. (2011). For this purpose, condition testing may not be limited to ordinarily checking the presence of certain entities, but may also rely on complex mathematical expressions – possibly based on attributes of entities – to describe certain spatial phenomena or arbitrary reaction rates, for instance. The general usefulness of such constraints and implications for the expressiveness of π-calculus-based languages has been demonstrated by John (2010). Moreover, the combination of flexible constraints with an activity scanning approach seems to constitute a distinguished applicability for modeling biochemical systems.

Henzinger et al. (2009) have compared diverse formalisms for specifying Markovian population models – i.e., models that are following a similar mathematical semantics as ML-Rules does – with respect to different properties, such as succinctness, compositionality, and expressive power. Thereby, all investigated languages are relatively simple formalisms lacking the concept of attributes. The study reveals that in comparison to matrix descriptions, stochastic Petri nets, stoichiometric equations, and stochastic process algebras, the language of *guarded commands* (Dijkstra, 1975) – a formalism exhibiting all characteristics of the activity scanning paradigm and allowing for arbitrary constraints – provides the most appropriate choice for describing such models. In a talk[2] in 2011, Tom Henzinger summarized the results of this comparative study as follows: *"... and the winner is: Guarded Commands!"*

Taken all together, rule-based modeling does not seem to conflict with any of the identified requirements for accessible multilevel modeling. Quite the contrary: a rule-based approach inherently enables a language design that meets many of them, namely a reaction-centric modeling paradigm, compositionality, succinctness, variable structures, and n-ary reactions. Moreover, the approach permits to support arbitrary rate kinetics and flexible constraints on reactions. Last but not least, the approach is generally considered to be user-

---

[2]Tom Henzinger. Syntax Matters. *Toward Systems Biology*, Workshop, May 30 - June 1, 2011, Grenoble, France.
Abstract: http://www-tsb-workshop.imag.fr/abstract_henzinger.html
Slides: http://www-tsb-workshop.imag.fr/slides/TSB_2011_Henzinger.ppt

friendly, since respective models are *"modular, maintainable, easy to modify, easy to implement, and easy to understand."* (Balci, 1988, p. 291). Although especially the maintainability property might not always hold true, as maintenance of truly large models might require additional means for black-box composition based on clearly defined interface descriptions (Röhl, 2008), a rule-based language design nonetheless seems to denote a suitable starting point toward accessible multilevel modeling.

6.2.3   Multisets of Attributed Entities

Like many other rule-based approaches, e.g., BNGL (Faeder et al., 2005, 2009), the κ-calculus (Danos et al., 2007a, 2009), BIOCHAM (Chabrier-Rivier et al., 2005; Fages and Soliman, 2008), LBS (Pedersen and Plotkin, 2010), or React(C) (John et al., 2011), ML-Rules employs also an *attributed* language design. That means, model entities are described in terms of *structured objects* that may be equipped with different states (see also Chapter 4.2). In ML-Rules, entities are called *species* and denote the basic building blocks of a model. As we will see later, a species may represent any object of interest, such as a small chemical, a protein, a membrane-bound compartment, or even an entire cell.

Each species has a name $X \in Names$, where $Names$ is the set of all names. A name X has a fixed arity $ar(X) \in \mathbb{N}_0$ that specifies the number of attributes of respective species. For the sake of simplicity of the language, attributes are not typed. Furthermore, they are not restricted to a predefined (i.e., finite) set of values and may thus be of any kind of numerical value or string of character symbols. By convention throughout the thesis, the name of a species is given in a bold font and attributes are written in a monospaced typeface embraced by a pair of parentheses behind the species name. Multiple attributes are separated by a comma and parentheses are omitted if $ar(X) = 0$, i.e., if species X has assigned no attributes.

For example, **A**(cyt, F) describes a protein species A that has assigned two attributes, the first describing the molecule's compartmental location (the cytoplasm in this case) and the second one the state of a binding site, i.e., whether the protein is free or has bound another molecule at this site. Similarly, **Cell**(1.38, G1) describes also a species with an arity of 2. In this case however, the species represents a cell entity whose attributes are representing its size and the cell cycle phase. **B** and **C**(nuc) are examples for species with $ar = 0$ and 1 respectively. Please notice, since the arity of a species name is fixed and may therefore not vary between species with identical names, the usage of **A**(cyt, F) and **A**(cyt) within the same model would be invalid. Also, each defined combination of species name and attribute values denotes a distinct species, which means that although **A**(cyt, F) and **A**(nuc, F) share the same name, they represent two different kinds of species.

Multiple species of the same as well as different kinds can be subsumed into multisets. According to homogeneous mixtures in chemistry, a multiset of species is called *solution*. A solution S can be either a single species, a composition of multiple sub-solutions, or an empty set $\emptyset$. The '+' is the delimiter symbol for composing multiple solutions. We write $n\,X$ with $n \in \mathbb{N}$ to refer to a solution which is composed of n identical copies of X, however, n is omitted if $n = 1$. For example,

$$S = 2\,\mathbf{A}(\mathtt{cyt}, \mathtt{F}) + 4\,\mathbf{A}(\mathtt{nuc}, \mathtt{F}) + \mathbf{C}(\mathtt{nuc})$$

describes a solution consisting of three different kinds of species with an amount of 2, 4, and 1 respectively.

6.2.4 Rule Schema Instantiation and Pattern Matching

Dynamic state changes of an ML-Rules model are described by reaction rules. Given a defined solution, a rule determines how this solution will change when the rule fires, i.e., a rule specifies the removal and addition of certain species from and to a given solution respectively. In other words, when firing, a rule substitutes a reactant solution S by a product solution S':

$$S \to S'\,.$$

This way, ML-Rules employs a classical reaction-centric modeling perception and the general rule syntax – which follows the notation of chemical reaction equations – can be similarly found in other rule-based approaches, e.g., BNGL and BIOCHAM. For simplicity reasons, ML-Rules only allows for specifying unidirectional rules. Thus, in order to model a reversible reaction, two complementary rules need to be defined. However, it would be straightforward to extend the syntax to reversible reaction rules like it is supported by BNGL, for instance.

A standard feature of rule-based languages is to specify rules in terms of rule schemata. That means, a rule may consist of schematic patterns to encode for various contexts and this is why rule-based languages are typically considered to facilitate succinct model descriptions. In the application field of systems biology, rule-based languages allow for specifying schematic rules typically by defining *reactant patterns*, such that certain attributes of entities taking part in a reaction are left unspecified if they are of no interest for this particular reaction. Therefore, a "don't care, don't write" approach is utilized by many languages, e.g., BNGL and the κ-calculus, where attributes of entities can simply be left away, while others – i.e., those attributes that are of relevance for this particular reaction – can be precisely specified by referencing attribute names in combination with a defined state or value. Attributes in ML-Rules, by contrast, are identified by their position rather than a name, which is similar to React(C) (John et al., 2011), for instance. Although this requires to enumerate each attribute of species, we do not need to define attribute names in this case while it is still possible to specify reactant patterns and thereby rule schemata, since attributes can be bound to variables serving as wildcards.

Let us take an example. Free molecules of protein A may degrade, however, the degradation reaction may take place within different cellular compartments. Let us assume that $ar(\mathbf{A}) = 2$, where the first attribute describes the compartmental location and the attribute at position two the state of a protein binding site. Without schematic rules, we would need to specify one distinct rule for each possible location of $\mathbf{A}$, which can be tedious and error-prone if we have many different compartments. Moreover, as the model's state space might be infinite due to not restricting attributes to a predefined set of potential values, it might be impossible to define each potential reaction in advance. Therefore, instead of specifying multiple rules with defined reactant species, of which each comprises a different value at attribute position one, we define a reactant pattern by inserting a variable l for the location attribute (F denotes a "free" binding site):

$$\mathbf{A}(l, \mathrm{F}) \to \emptyset \ .$$

A mapping of this rule schema to the above solution S would then evaluate to the following two rule instantiations

$$\mathbf{A}(\mathtt{cyt}, \mathrm{F}) \to \emptyset$$

$$\mathbf{A}(\mathtt{nuc}, \mathrm{F}) \to \emptyset$$

where l has been bound to values $\mathtt{cyt}$ and $\mathtt{nuc}$ respectively.

Please notice, the name of such a variable is only locally known, i.e., variables are private such that they are restricted to the scope of a single rule schema. However, by using a variable name repeatedly within the same rule schema it is possible to constrain bindings of multiple species having identical attribute values, e.g., to describe n-ary reactions that require each reactant species to be located within the same compartment. In addition, besides such basic examples, rule schemata can be also specified by using expressions based on already bound variables. For instance, a reactant pattern

$$\mathbf{A}(x) + \mathbf{A}(2x) \to \ldots$$

matches every solution where at least two species with name $\mathbf{A}$ exist, one of which attribute's value is exactly twice the attribute's value of the other one. Similarly, expressions based on bound variables can be also used to specify attribute values assigned to product species, e.g., the rule

$$\mathbf{A}(x) \to \mathbf{B}(2x)$$

applied to a solution $S = \mathbf{A}(2) + \mathbf{B}(3)$ would lead to a product solution $S' = \mathbf{B}(3) + \mathbf{B}(4)$.

Please note, in principle the language concept of ML-Rules allows for any kind of expression. However, in practice the set of valid operators and functions depends on the concrete implementation of the language, as will be also explained in Chapter 6.3. However, we consider a small set of frequently applied operators and functions being essential. Therefore, each realization should support at least the four basic arithmetic operations (addition, subtraction, multiplication, division), square roots, exponentiation, binomial coefficients, exponential function, logarithmic functions, conditional expressions (if-then-else), equality and inequality relations (equal, less than, greater than, less than or equal, greater than or equal), and the three basic Boolean operations (and, or, not).

6.2.5 Species Bonding

Besides the general approach of rule schema instantiation, another common method for reducing the size of a model is to support some notion of linkage between entities, e.g., to represent noncovalent bonds within protein complexes. By doing so, individual subunits of a protein complex with possibly many different modifications can be preserved instead of specifying numerous species names, each of which would reflect a different combination of subunit states and bindings.

In ML-Rules, links between species are represented by attributes, i.e., unique names or identifiers the bound species are assigned with. Entirely new values can be created with the help of an operator ν. For example, rule

$$\mathbf{A}(F) + \mathbf{B}(F) \rightarrow (\nu x)\,\mathbf{A}(x) + \mathbf{B}(x)$$

describes the binding between two species $\mathbf{A}$ and $\mathbf{B}$, where the attribute F denotes a free binding site and νx creates a fingerprint-like unique value which does not already occur in the current model state. It is assigned to the products on the right-hand side of the rule via variable x, i.e., in an instantiation of the rule schema, x is replaced by a newly created unique value which serves as an identifier for this particular binding.

This method for representing linkage of species is identical to private channel names in the π-calculus and allows to model molecular complexes similar as can be done with established rule-based languages that have explicit notions of complexation, e.g., BNGL or the κ-calculus. Moreover, once created, unique values can be used in a highly flexible manner, e.g., to describe bonds across the boundaries of hierarchical levels (the concept of multiple levels will be introduced below) or that are shared by more than two binding partners (i.e., hypergraphs), like it is also supported by Bigraphs and React(C). In addition, certain species may be also marked with unique identifiers to observe or track the dynamics of individual entities.

However, it should be noted that the approach has also some drawbacks compared to explicit notions of molecular bindings. First of all, without profound knowledge or decent annotation of the model, it might be difficult to find out whether certain attributes of species represent binding sites or not. The approach would also allow resetting just one species to its unbound state while its former binding partner remains unchanged. Therefore, the modeler carries the responsibility for a correct model.

6.2.6 Kinetic Rates and Dynamic Constraints

As the general mathematical semantics of ML-Rules is based on continuous-time Markov chains (CTMCs), each rule consists not only of a reactant and a product solution, but is also assigned a stochastic kinetic rate $r \in \mathbb{R}_0^+$:

$$S \xrightarrow{r} S' \ .$$

That means, given a certain model state, a rule schema does not only define in which way the state of the model changes when firing but also how frequently this will happen. Thereby, like in any other CTMC-based approach, the process of firing is stochastic and the waiting time for firing follows a negative exponential distribution. In other words, the higher the rate of a concrete rule instantiation (propensity), the more likely its firing will be at a point in time that is also calculated according to this rate.

The kinetic rate of a rule can be a constant numerical value, for example, to describe a chemical reaction with constant speed, i.e., a zero-order reaction whose speed does not depend on the amount of any species. However, reaction rates of biochemical systems typically vary over time, as they depend on the varying amounts of one or more reactant species. Therefore, the rates or propensities of reaction rules need to be dynamically adjusted according to the current state of the model.

Many languages and tools for modeling biochemical systems, e.g., simple stoichiometric equations, the stochastic π-calculus, and the κ-calculus, implicitly assume that the kinetics of each reaction follows the law of mass action.

The firing rate is thus automatically adjusted and the according mathematical formula does not need to be specified by hand. This way, modeling is less error-prone and models become also more succinct, as solely the specific reaction rate constant k (also known as rate coefficient) needs to be specified. However, in this case, modeling is limited to elementary reactions. Abstractions like the Michaelis-Menten approximation for enzymatic reactions or Hill kinetics for describing cooperativity are not possible, although we have already learned earlier that such abstractions are often required due to missing data or to reduce the complexity of models. Supporting arbitrary rate kinetics thus often denotes an important feature for modeling in systems biology and that is why there is an increasingly large number of modeling approaches that do not necessarily rely on the law of mass action.

A support for arbitrary kinetics is also one of the essential requirements we have identified for multilevel modeling of biological systems in general. The kinetic laws in ML-Rules are therefore explicitly specified by using mathematical expressions as reaction constraints. Thereby, similar to the attributed π-calculus, any kind of expression is allowed as long as evaluating to a nonnegative numerical value.

Species identifiers are used to refer to the amounts of certain species in a given solution. The notation $\mathbf{A}(\dots)^a$ assigns a species identifier a to reactant $\mathbf{A}(\dots)$ and we write $\#a$ to access the amount of a particular species that has been bound during rule instantiation. For example, assuming mass-action kinetics, the rate of a first-order reaction $\mathbf{A}(x) \to \mathbf{B}$ with rate coefficient k is correctly described as follows:

$$\mathbf{A}(x)^a \xrightarrow{k \cdot \#a} \mathbf{B} \ .$$

A mapping to solution $S = 2\,\mathbf{A}(1) + 4\,\mathbf{A}(2)$ then evaluates to two different rule instantiations

$$\mathbf{A}(1) \xrightarrow{k \cdot 2} \mathbf{B}$$

$$\mathbf{A}(2) \xrightarrow{k \cdot 4} \mathbf{B}$$

each with a different propensity due to the different amounts of species $\mathbf{A}(1)$ and $\mathbf{A}(2)$ in solution S.

Besides such basic kinetics, expressions can be also used to control the dynamics of models more flexibly. The idea is basically the same like in colored Petri nets, the attributed π-calculus, or React(C), where mathematical expressions allow for specifying rates based on attributes. For example, the rate of the following rule decreases with increasing values of attribute x:

$$\mathbf{A}(x)^a \xrightarrow{\#a/x} \mathbf{B} \ .$$

Moreover, conditional expressions are also supported, e.g., the preceding rule can be constrained to only fire if the amount of $\mathbf{A}(x)$ exceeds a certain threshold value *thres*:

$$\mathbf{A}(x)^a \xrightarrow{\text{if } \#a > thres \text{ then } k \cdot \#a \text{ else } 0} \mathbf{B} \ .$$

If the amount of $\mathbf{A}(x)$ does not exceed *thres*, the conditional "if-then-else" expression evaluates to a zero kinetic rate, which means that the rule will not fire at all as long as the condition ($\#a > thres$) remains unfulfilled.

As such conditional reaction constraints easily result in longish rate expressions and – as shown later – are frequently applied constructs for modeling interlevel causation, an extra notation is used to improve the readability. Therefore, instead of a complex rate definition consisting of an expression "if e then r else 0", we write the conditional expression e below the arrow that is assigned the basic kinetic rate r:

$$S \xrightarrow[e]{r} S' \quad \triangleq \quad S \xrightarrow{\text{if } e \text{ then } r \text{ else } 0} S' \ .$$

The preceding example now looks as follows:

$$\mathbf{A}(x)^a \xrightarrow[\#a > thres]{k \cdot \#a} \mathbf{B} \ .$$

6.2.7 Nested Species Multisets

The above features such as attributed species, rule schemata, arbitrary rate laws, and flexible constraints, are neither new nor sufficient for accessible multilevel modeling, but all together they play important roles in supporting multilevel modeling of biological systems, as has been extensively discussed earlier in this thesis.

However, a truly salient feature of multilevel modeling are hierarchies, i.e., the capability to structure models in a hierarchically nested manner. Hierarchical structuring facilitates modeling of complex biological systems by defining them in terms of their components and the interactions that exist between them. Thereby, hierarchies may help to structure the knowledge about a given system, e.g., by explicitly describing the multiple nested reaction compartments that can be found in biological systems, such as cells, organelles, and vesicles.

To address the need for hierarchical model structures, the concept of *nested species* is incorporated into ML-Rules. That means, species may not only be characterized by names and their attributes, but also by the context (or environment) and a potentially enclosed sub-solution of further species. The approach is pretty much inspired by other hierarchical modeling approaches, such as BioAmbients, DEVS, or Bigraphs, where entities may have one unique superior entity and possibly several constituents. The model structure can thus be represented graphically by a disjoint union of rooted trees, i.e., a forest. In addition, since ML-Rules pursues a population-based approach, identical species that are indistinguishable from each other are aggregated to populations of certain species. However, both the parent (context) as well as children (content or sub-solution) are thereby taken into account. Let us take an example: Solution

$$S = \mathbf{A}\big[\overbrace{\mathbf{B} + 4\,\mathbf{B}[\underbrace{\mathbf{C}}]\,}^{S_{\mathbf{A}}}\big] + \mathbf{B}[\underbrace{3\,\mathbf{C}}] + \mathbf{B}[\underbrace{\mathbf{C}}]$$
$${\scriptstyle S_{\mathbf{B}_1}} \qquad {\scriptstyle S_{\mathbf{B}_2}} \qquad {\scriptstyle S_{\mathbf{B}_3}}$$

consists of three species, one $\mathbf{A}$ and two $\mathbf{B}$, each of which contains a solution with further species on its own. The square bracket syntax for representing nested structures thereby follows the BioAmbients notation. Species $\mathbf{A}$ consists of a sub-solution $S_{\mathbf{A}} = \mathbf{B} + 4\,\mathbf{B}[\mathbf{C}]$, i.e., two species $\mathbf{B}$: an atomic one that is lacking an own sub-solution and a species $\mathbf{B}$ comprising a sub-solution $S_{\mathbf{B}_1} = \mathbf{C}$. Species $\mathbf{A}$ encloses four identical copies of this nested species so that they are aggregated to a population of the respective amount. The other two species $\mathbf{B}$ residing at the same level like $\mathbf{A}$ contain both a sub-solution consisting of an atomic species $\mathbf{C}$. However, sub-solution $S_{\mathbf{B}_2}$ consists of a population with an amount of 3 $\mathbf{C}$ species, while $S_{\mathbf{B}_3}$ consists of only one $\mathbf{C}$. Both species $\mathbf{B}$

thus denote different kinds of nested species and are therefore not aggregated. The corresponding graph representation looks as follows:

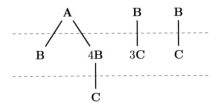

Please notice, nested species may still have assigned attributes. Attributes and the enclosed solution are embraced by different kinds of brackets, e.g., $A(0)[S_A]$ denotes the existence of an attribute 0 for the above species A. The capability to assign attributes to nested species allows us to equip each hierarchical level with an own state that is not solely determined by its enclosed sub-solution (reductionistic view), but may rely on dynamics that is independent from lower levels. Hence, in this respect the approach is similar to hierarchical model structures in ML-DEVS (Uhrmacher et al., 2007) or EMSY (Uhrmacher, 1995), for instance. Supporting states and behavior at any level is of particular interest for multilevel modeling, as it allows for modeling dynamic behavior similar to observations made at different levels of organization. Moreover, it denotes the basis for describing upward and downward causation in a natural and straight manner.

Unlike the entities of modular and object-oriented reactive systems approaches, nested species in ML-Rules do not enclose the specification of a model's dynamics. In DEVS and EMSY, for instance, the behavioral rules are part of the entities and clearly defined interfaces allow for an interaction with the environment (Uhrmacher, 1993; Uhrmacher and Zeigler, 1996). ML-Rules – by contrast – employs a different approach, since species contain nothing else than other species and the nested hierarchy thus only describes the structure of a model, but not its dynamics. The latter is globally specified by a set of rule definitions. Therefore, in this respect ML-Rules is very similar to the language of Reactive Bigraphs, which also separates entities and the nested model structure from the definitions of dynamic behavior.

6.2.8   Multilevel Rule Schemata

A nested hierarchical model structure enables to reduce model complexity not only by specifying rule schemata as described above, i.e., by specifying reactant patterns, in which attributes are bound to variables and thereby different attribute values lead to different rule instantiations. The number of rules may be also reduced by applying a single rule to multiple sub-solutions being part of the hierarchical model structure, so that reactants can be matched within different solutions at different levels. Thereby, the reactants may be even enclosed by different species types (cf. Figure 6.1), such that, for example, it is possible to describe a reaction taking place within various cellular compartments by specifying a single rule.

However, pattern matching within a population-based hierarchy of nested species has an important consequence for the semantics. When applying rule schemata to diverse matched solutions and calculating the propensity of according rule instantiations, the context of each application needs to be taken into account, which is given by the amount of species at higher levels. That means, the propensity of a rule instantiation needs to be adjusted according to the whole hierarchy above, as a reaction or state transition is more likely to happen the more solutions exist it could potentially take place in. Figure 6.2 illustrates the general procedure with the help of a simple example.

What makes no sense in a hierarchical setting is to allow for an empty set of reactant species, e.g., to describe a zero-order synthesis of certain molecules. As rules like

$$\emptyset \xrightarrow{k} A$$

do not require to fulfill any condition in order to fire, the process of pattern matching would result in one rule instantiation per sub-solution. That means, each newly synthesized species also serves as a potential location within which the next reaction may take place. Consequently, after a few firings we would get something like $A[A[A[A[\dots]]]]$. Hence, in order to describe meaningful zero-order reactions, the specification of a *context* is mandatory.

Context means the specification of defined locations at which a certain reaction may take place. Therefore, but also to let different levels of a hierarchical model interact with each other to describe upward and downward causation,

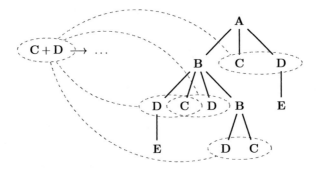

Figure 6.1: Reactant pattern matching within hierarchical model structure. One rule schema may be applied to multiple sub-solutions.

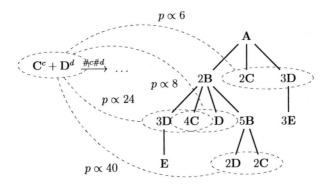

Figure 6.2: Multilevel rule schema instantiation. The actual propensity p of an instantiation depends not only on the specified rate law and quantities of matched reactants, but also on their context within the nested model structure.

rule schemata in ML-Rules may involve nested reactant and product species. Such *multilevel rules* look pretty much the same like rules that are independent from a defined context and that are operating on a single level only, except of the fact that multilevel rules comprise nested species. For example,

$$\text{Cell}[\,]^c \xrightarrow{k \cdot \# c} \text{Cell}[A]$$

specifies a contextual rule schema that restricts the above zero-order reaction to take place only within a **Cell** compartment.

Of course, specifying nested reactants and attributed species is also possible. Thereby, nested species with arbitrarily many levels are allowed, since interlevel causation in biological systems is not necessarily limited to neighboring levels, in particular when abstracting from certain detailed aspects. For example, transcription of a gene, translocation of produced mRNA from the nucleus into the cytoplasm compartment, translation of mRNA into protein, and finally the secretion of protein into the extracellular medium may be described as one abstract process spanning multiple organizational levels. The enumeration of according rule instantiations works, in principle, similar to what has been explained earlier. The only difference is that the process of instantiating multilevel rule schemata requires to take multiple levels and solutions into account when matching reactants. For instance, the rule

$$\mathbf{A}(0)[\mathbf{B}] \rightarrow \mathbf{A}(1)[\mathbf{B}]$$

describes a reaction from $\mathbf{A}(0)$ to $\mathbf{A}(1)$ under the condition that $\mathbf{A}$ encloses at least one species $\mathbf{B}$ (the kinetic rate expression is omitted).

Please note, the reactant pattern $\mathbf{A}(0)[\mathbf{B}]$ also matches species where $\mathbf{A}(0)$ contains further species in addition to the explicitly mentioned $\mathbf{B}$, no matter which and how many. That means, we do not need to specify the entire content, as otherwise this would easily result in a very large number of cluttered rule definitions and in many situations it would even be impossible due to dynamically changing sub-solutions. However, in the above example, the remainder of the potentially existing sub-solution would get lost when the rule fires, since the product species on the right-hand side contains just exactly one $\mathbf{B}$. The same holds true for a potential sub-solution of $\mathbf{B}$: the reactant pattern matches every species where a **B** is part of a sub-solution of an **A**(0), but it says nothing about a sub-solution of **B**. Hence, if the reactant species **B** contains further species, they would get lost as well.

For understanding the general functionality here, it is important to bring the concept to mind that ML-Rules does not assume any implicit insertion or preservation of species within a product solution, i.e., the products that are inserted into the system when a rule fires are exactly those that have been specified on the right-hand side of the rule. To put it in other words, nested product species contain only those sub-solutions that have been explicitly specified. The reason for this lies in the fact that – strictly speaking – ML-Rules does not include a concept of species *modification*. Instead, the general rule firing semantics directs a *substitution* of species, i.e., a removal of reactant species while at the same time fresh products are put into the system. If certain unspecified attributes of reactants shall thereby be preserved, they need to be bound to variables that can then be used to assign the according values to product species (cf. Section 6.2.4). The preservation of sub-solutions works rather similar.

To prevent from the loss of a potentially existing but unspecified multiset of species in the above reaction, the remainder of the solution enclosed by **A** needs to be bound to a special variable. We write x? for specifying such a variable that binds a remainder-solution, where x denotes the variable name. By reinserting this variable into the product species, the entire remainder-solution bound on the left-hand side can be preserved without the need to explicitly specify its exact content:

$$\mathbf{A}(0)[\mathbf{B} + x?] \rightarrow \mathbf{A}(1)[\mathbf{B} + x?] \ .$$

By binding more than one remainder-solution it is also possible to preserve the content of multiple species, e.g., firing of the rule

$$\mathbf{A}(0)[\mathbf{B}[x_\mathrm{B}?] + x_\mathrm{A}?] \rightarrow \mathbf{A}(1)[\mathbf{B}[x_\mathrm{B}?] + x_\mathrm{A}?]$$

would preserve the entire content of both species **A**(0) and its enclosed **B**.

At first sight this approach for preserving unspecified content might seem unnecessarily complicated, as one could alternatively also imagine to simply assume an implicit preservation (cf. the *frame problem* in artificial intelligence formulated by McCarthy and Hayes, 1969). However, in this case additional language constructs are needed to describe processes where the deletion of sub-solutions is intended. Moreover, the presented approach of explicitly binding remainder-solutions has the advantage of making the semantics generally more clear and transparent without the need to cover various special cases. For example, let us assume a reaction $A + B \rightarrow C + D$, where two species are simultaneously converted into two other ones and both reactant species enclose an own sub-solution. A semantics assuming implicit preservation of sub-solutions would require an automated mapping from reactants to products, which would be problematic in cases where the numbers of reactants and products differ, e.g., two reactant and three product species. It would also imply that the order of species matters, so that

$$A[\,] + B[\,] \rightarrow C[\,] + D[\,] \quad \text{and}$$
$$A[\,] + B[\,] \rightarrow D[\,] + C[\,]$$

would describe different reactions, where the product $\mathbf{C}$ encloses in the first case the content of reactant $\mathbf{A}$ and in the second case that of reactant $\mathbf{B}$. The decision against implicit preservation, by contrast, ensures transparently full control of what happens with bound solutions and thereby also facilitates the description of multilevel phenomena like migration and fusion of compartments.

In Chapter 8, practical examples from biology illustrate the capabilities of the presented multilevel approach more deeply and also show how to describe upward and downward causation in realistic case studies.

6.3 Realization

A prototypical implementation of the presented ML-Rules modeling concept has been realized within the multi-purpose modeling and simulation framework JAMES II (Himmelspach and Uhrmacher, 2007) and its source code has been made available under http://www.jamesii.org/. The plugin-based JAMES II framework consists of a *core* and a rich set of additional *plugins*,

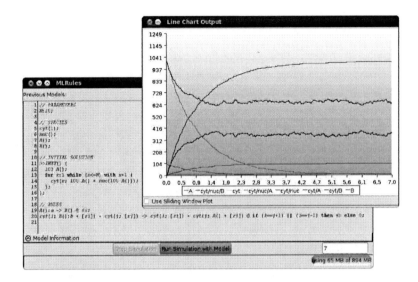

Figure 6.3: ML-Rules model editor and line chart visualization.

such as diverse alternatives for data storage, random number generation, and event queues. The core defines a basic set of plugin types and plugins needed to run *in silico* experiments. Also part of the core, the registry is responsible for managing plugin types and plugins, and the experimentation layer carries out repeatable and reusable simulation experiments, e.g., simple (parallel) simulation runs, parameter scans, optimizations, and sensitivity analyses (Himmelspach et al., 2008; Ewald et al., 2010).

For ML-Rules, a set of additional plugins has been implemented: an editor that allows to create and edit ML-Rules models and supports syntax highlighting based on syntactical and semantical consistency checks (Figure 6.3), plugins for model reading and writing, and a simulator which is based on the *Direct Reaction Method* of Gillespie (1977) and thus implements an exact stochastic simulation algorithm (SSA). In addition, a solution based on an SQL-like query language for flexible instrumentation and observation of ML-Rules models has been developed (Helms et al., 2012).

JAMES II and the ML-Rules specific plugins are written in Java. Therefore, based on Java reflection, functions provided by Java can be used within

147

expressions, e.g., those defined by the class java.lang.Math[3]. A list of the most important mathematical operators and functions for specifying expressions in ML-Rules is given in Table 6.1. The integration of a library of user-defined functions is not yet supported but may be realized in the future.

The syntax of the ML-Rules implementation has undergone slight adaptations compared to the aforementioned abstract notations, so that models can be specified by using the standard ASCII[4] character encoding scheme. In the following, crucial aspects of the concrete syntax will be explained with the help of syntax diagrams and grammar snippets (the complete grammar in EBNF[5] notation can be found in Appendix A) as well as simple examples.

Finally, a brief description of the simulation algorithm will be presented.

6.3.1 Concrete Syntax of Model Specification

An ML-Rules model consists of four distinct parts to encode the different types of essential information: An optional set of constants, a set of species definitions, the initial solution, and a set of rules.

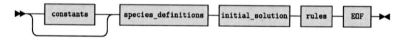

```
definitions
      ::= constants? species_definitions initial_solution rules EOF
```

Names of species, constants, and variables may generally consist of the following set of character symbols {A-Z, a-z, 0-9, _}. Values may be numerical (INT and FLOAT), strings of arbitrary character symbols enclosed by single quotes (STRING), or a Boolean (true/false). ν-binders, i.e., entirely unique values, are generated by "$n", where n is a local name that allows to assign the same value to more than one species at the same time (see example on page 152). Please note, the ML-Rules language concept does not include a type system. However, in a concrete realization, distinguishing between different types of data facilitates syntactical and semantical consistency checks.

---

[3]http://docs.oracle.com/javase/6/docs/api/java/lang/Math.html
[4]ANSI X3.4-1968, http://tools.ietf.org/html/rfc20
[5]W3C Grammar Notation, http://www.w3.org/TR/xquery/#EBNFNotation

Table 6.1: Basic set of operators and functions for specifying mathematical expressions in ML-Rules. Restrictions regarding the data type of evaluated expressions e are given in the right column.

Operation/Function	Syntax	Restrictions
Addition	e_1 + e_2	$e_1, e_2 \in \mathbb{R}$
Subtraction	e_1 - e_2	$e_1, e_2 \in \mathbb{R}$
Multiplication	e_1 * e_2	$e_1, e_2 \in \mathbb{R}$
Division	e_1 / e_2	$e_1 \in \mathbb{R}, e_2 \in \mathbb{R} \backslash 0$
Exponentiation	e_1 ^ e_2	$e_1, e_2 \in \mathbb{R}$
Square root	`sqrt`(e)	$e \in \mathbb{R}$
Binomial coefficient	`binom`(e_1, e_2)	$e_1, e_2 \in \mathbb{N}$
Exponential function	`exp`(e)	$e \in \mathbb{R}$
Natural logarithm	`log`(e)	$e \in \mathbb{R}$
Logarithm of base 10	`log10`(e)	$e \in \mathbb{R}$
Conditional expression	`if` e_1 `then` e_2 `else` e_3	$e_1 \in \{\text{true}, \text{false}\}$
Equal	e_1 == e_2	
Not equal	e_1 != e_2	
Less than	e_1 < e_2	$e_1, e_2 \in \mathbb{R}$
Greater than	e_1 > e_2	$e_1, e_2 \in \mathbb{R}$
Less than or equal	e_1 <= e_2	$e_1, e_2 \in \mathbb{R}$
Greater than or equal	e_1 >= e_2	$e_1, e_2 \in \mathbb{R}$
Logical AND	e_1 && e_2	$e_1, e_2 \in \{\text{true}, \text{false}\}$
Logical OR	e_1 || e_2	$e_1, e_2 \in \{\text{true}, \text{false}\}$
Logical NOT	! e	$e \in \{\text{true}, \text{false}\}$

ML-Rules model specifications may be annotated by comments that will not be analyzed after parsing:

```
// this is a comment on a single line
/* this is a rather large comment and may span multiple
   lines */
```

Constants

Constants are defined at the beginning of a model specification and they serve as globally accessible model parameters. Each constant definition starts with a globally unique name followed by a colon and an expression to define its value.

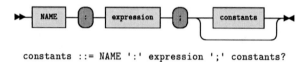

```
constants ::= NAME ':' expression ';' constants?
```

Expressions may thereby also depend on other constants, as is shown in the following example:

```
k1:0.027;
k2:10*k1;
state_P:'phosphorylated';
ON:true;
OFF:!ON;
```

Species definitions

In the set of species definitions the general properties of potentially occurring species are defined. Therefore, names of species and their respective arity are given, i.e., the attribute number of each species. The arity of species given a certain name is defined by a following non-negative integer value within parenthesis. An integer value may be omitted if the arity is zero.

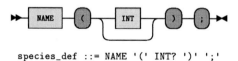

```
species_def ::= NAME '(' INT? ')' ';'
```

The listing below shows diverse examples of species definitions, each comprising a different number of attributes.

```
Nucleus(1);
p53();
EGFreceptor(5);
protein_X(2);
```

Initial solution

The initial state of the model is defined by its initial solution, i.e., a possibly nested multiset of species. Distinct elements are thereby separated from each other by the "+" symbol. A keyword (>>INIT) and square brackets are used to mark the beginning and end of its definition.

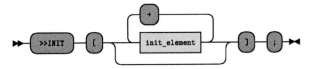

```
initial_solution
      ::= '>>INIT' '[' (init_element ( '+' init_element )*)? ']' ';'
```

Arbitrarily nested species containing sub-solutions of other species may be defined in the initial solution. Species quantities may be specified with the help of mathematical expressions and the same holds true for attribute assignments. *For loop* statements can be used to repeatedly define numerous similar species, for instance, like it is shown in the following example of an initial solution:

```
>>INIT[
        10000 protA(parameter) +
        parameter/100 ProteinB +
        Nucleus[
                gene('F') +
                20 TF('F','unphos')
                ] +
        for x:1 while (x<=100) with x+1 [ C(x) ]
        ];
```

Rules

The next and final part of a model specification defines a set of rules, which determines the model's dynamics given a certain state. Each rule is thereby generally structured as follows:

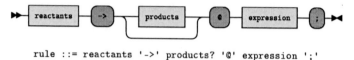

```
rule ::= reactants '->' products? '@' expression ';'
```

A similar syntax like for the initial solution is used to specify the sets of reactants and products of a rule. However, local variables may be inserted for specifying species patterns and binding remainder-solutions. In addition, reactants may have assigned a species identifier to retrieve a species' quantity, for instance. The set of products is optional. Like in KaSim[6] – a realization of the κ-calculus – the kinetic rate expression is specified at the end of a rule and follows an "@" symbol.

Some syntactically correct example rules are given below.

```
A:a + B:b -> AB @ k * #a * #b;
C(0):c + D(0):d -> C($bond) + D($bond) @ k * #c * #d;
Nucleus[A:a + sol?] -> Nucleus[sol?] + A @ #a * k_n2c;
2 A:a -> AA @ binom(#a,2);
#a A:a -> @ if (#a>100) then k_delete_all else 0;
```

6.3.2 Simulation Algorithm

A basic pseudocode description of the realized simulation algorithm is presented in Figure 6.4. Calculating the next simulation step requires three different sets of information: The invariable set of rule schemata given by *Rules*, the current state of the model (*State*), and the set of constant model parameters (*Consts*).

MatchReactants selects all matching reactants R_s of the selected rule schema s. To find all matching species and since reactants may be nested, an extensive search through the hierarchy of the model's *State* is required. Coarsely

---

[6] κ-calculus implementation KaSim: http://kappalanguage.org/

Require: *Rules, State, Consts*

1: **for all** $s \in Rules$ **do**
2:     $R_s \leftarrow MatchReactants(s, State)$
3:     $Instances \leftarrow Instances \mid CreateInstances(s, R_s, State, Consts)$
4: **end for**

5: **for all** $i \in Instances$ **do**
6:     $i_{\text{prop}} \leftarrow CalcPropensity(i)$
7:     $Reactions \leftarrow Reactions \mid CreateReactions(i, i_{\text{prop}})$
8: **end for**

9: $reaction \leftarrow SSA(Reactions)$
10: $Reactants \leftarrow CreateReactants(reaction)$
11: $Products \leftarrow CreateProducts(reaction)$

12: **for all** $r \in Reactants$ **do**
13:     $RemoveReactant(r, State)$
14: **end for**

15: **for all** $p \in Products$ **do**
16:     $PutProduct(p, State)$
17: **end for**

Figure 6.4: Basic simulation algorithm of the ML-Rules implementation.

50% of the overall simulation time goes into *MatchReactants* (data not shown). However, giving a complexity measure is rather difficult, as the effort for matching reactants strongly depends on the modeled system. A coarse estimate of the complexity for matching one reactant is $\mathcal{O}(n \cdot m^k)$, where n denotes the overall number of different species (i.e., nodes in the hierarchy graph), m denotes the average number of different species in one sub-solution (i.e., the average child node number of nodes having children), and k denotes the depth of nesting of the reactant. That means, the time or computation steps required for matching a non-nested reactant grows linearly with the total number of species n, i.e., with the overall number of nodes. If the reactant is specified within a context, however, the effort for matching this nested reactant is influenced by an additional factor depending on the size of sub-solutions and growing exponentially with the depth of nesting of the reactant.

Let us consider an example to outline the procedure of *MatchReactants*. Given a non-nested reactant pattern, e.g., **B**, requires to visit each node of the hierarchy graph once only to find all species matching this pattern. Matching the nested reactant pattern **A**[**B**+*r*?], by contrast, is solved by running through the entire graph to find all species **A** first. Afterwards, the child nodes of all matched **A**s need to be visited again to find all species **B** that are enclosed by a species **A**. In the example hierarchy below, for instance, the nodes marked by an asterisk are visited twice to match the encircled reactants and therefore matching nested reactants is computationally more expensive compared to non-nested ones. However, some optimizations are employed to restrict the search space for matching reactants, e.g., when matching the reactant pattern **C** + **A**[**B** + *r*?], sub-solutions that do not contain species **C** are not considered for searching the second reactant **A**[**B** + *r*?].

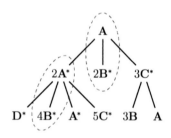

The second step after matching reactants is to instantiate the rule schema *s* by calculating the set of rule *Instances*. Afterwards, the propensity is calculated for each instance *i*. With *CalcPropensity* the propensity of each generated instance is calculated using the specified expression and taking the context of the matched solution into account. This means that the propensity is adjusted according to the amounts of possible contexts the matched solution is part of (cf. Section 6.2.8 and Figure 6.2). Please note, as the number of reactants that apply is part of the instance, calculating the propensity is only dependent on *i*. Also, all information needed is directly available for each product in *i*, such as bound attribute values or solutions that are used on the rule's product side. After calculating the set of potential *Reactions*, an SSA is invoked, e.g., the *Direct Reaction Method* of Gillespie (1977), and thereby the next *reaction* is determined.

The selected reaction is executed by removing reactants r and adding products p from respectively to the current *State*. Maintaining the consistency of populations when executing *RemoveReactant* and *PutProduct* is a crucial part during simulation and requires – given the nested species – special attention. This means, whenever a species x is removed from or added to a sub-solution S_{sub} from the overall solution S, the populations within S need to be updated accordingly. Sometimes it might not be sufficient to just decrease or increase the population value of x in S_{sub}, because by removing or adding x the superordinate species that encloses S_{sub} becomes a different species which means it might need to be split from the previous population it was attached to and needs to be merged with an already existing one. This actually has to be carried on upward the hierarchy until no splitting and merging is needed anymore. Let us take an example to illustrate this point. Given a solution

$$S = 2\,\mathbf{A}[2\,\mathbf{B}] + 2\,\mathbf{A}[3\,\mathbf{B}]$$

comprising of two different populations of a nested species $\mathbf{A}[\dots]$, removing one $\mathbf{B}$ from $\mathbf{A}[3\,\mathbf{B}]$ would first lead to a split of the population of $\mathbf{A}[3\,\mathbf{B}]$ and result in the temporary solution

$$S^* = 2\,\mathbf{A}[2\,\mathbf{B}] + \mathbf{A}[3\,\mathbf{B}] + \mathbf{A}[2\,\mathbf{B}]\ ,$$

where $\mathbf{A}[2\,\mathbf{B}]$ has to be merged with the already existing population of $2\,\mathbf{A}[2\,\mathbf{B}]$ to ensure consistency. So, the correct successional solution will be

$$S' = 3\,\mathbf{A}[2\,\mathbf{B}] + \mathbf{A}[3\,\mathbf{B}]\ .$$

Chapter 7

Syntactic Sugar and Meaningful Extensions

In this chapter, different extensions and syntactical refinements of the approach presented in Chapter 6 are discussed. Although the presented ideas have not yet been implemented as part of the prototypical realization, it is shown that they fit well into the general modeling concept of ML-Rules – e.g., by simple transformations from additional syntactic notations to basic ones – and that they denote valuable improvements of the language to further enhance its applicability and accessibility for multilevel modeling in systems biology.

7.1 Graphical Syntax

7.1.1 Background

It is widely acknowledged that graphical and diagrammatic notations may enhance the accessibility of modeling languages – in particular for non-experts and novice users – and, compared to textual notations, diagrams may make it more easy to comprehend the structure and interactions of a modeled system (see, e.g., Green et al., 1991; Paige et al., 2000; Alves et al., 2006; Phillips et al., 2006; Le Novère et al., 2009).

However, a graphical syntax is not necessarily of advantage, as has been shown by diverse comparative user studies (Green et al., 1991; Green and Petre, 1992; Moher et al., 1993). One of the main reasons for this observation

lies in the *secondary notation* of graphical languages, such as layout and ty-
pographic cues, which may cause serious confusion if not carefully designed.
Marian Petre (1995, p. 43) describes the problem as follows: *"The giddy inter-
twining connections of a too-complex or poorly designed boxes-and-lines-style
representation makes vivid the notion of 'spaghetti code'."* Also, the understandability of diagrams depends – similar to textual notations – on readership skills. That means, an accessible language should not call for exhaustively
studying the meanings of graphical notations beforehand but instead should be
rather intuitive by meeting as many as possible conventions the users expect,
e.g., an arrow is commonly understood to represent some kind of transition or
information flow in a certain direction and should thus be used accordingly. In
general, *"'Good' graphics usually means linking perceptual cues to important
information, which means both identifying and capturing what is important,
and guiding the reader with appropriate cues."* (Petre, 1995, p. 43). Therefore,
an appropriately designed graphical notation might significantly support users
in their tasks, but one needs to be also aware of the problems of overloading
diagrams with too much information and introducing unexpected meanings of
certain visual representations.

7.1.2   A Graphical Notation for ML-Rules

When we are looking at the textual syntax of ML-Rules as introduced in the
previous chapter, there is one notable point where a graphical notation might
be definitely useful to increase the readability: By using a textual notation,
the specification of nested species may involve multiple nested pairs of squared
brackets to distinguish the different sub-solutions being part of a hierarchy, i.e.,
which are both vertically as well as horizontally separated from each other.
Thereby, it might be difficult to perceive which opened bracket belongs to
which closed one and thus the hierarchical levels of certain species might be
unclear at first sight. The nested species below shall illustrate the problem.

$$A[B[C + 2\,D[E]] + B[2\,C + 3\,E]]$$

Here, a graphical notation may help to avoid confusion by too many brackets and to get a quick visual impression of the nested structure. Therefore,

inspired by Harel's Higraph notation (Harel, 1988) and thus similar to other visual and hierarchical modeling approaches like Statecharts and Bigraphs, a graphical representation of nested nodes can be used to facilitate the readability of hierarchical relationships. By drawing rectangular boxes instead of writing brackets, a transformation between the textual and graphical syntax is straightforward and clear:

$$\mathbf{X}[S_{\mathbf{X}}] \triangleq \boxed{\begin{array}{l} \mathbf{X} \\ S_{\mathbf{X}} \end{array}}$$

where $S_{\mathbf{X}}$ denotes the sub-solution that is enclosed by species $\mathbf{X}$. The more complex example from the beginning then looks as follows:

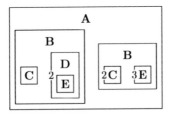

Please notice, the stoichiometric factors at the edge of some of the boxes are describing amounts of species populations greater than one.

In addition to the nested nodes representation, colors or grayish background shadings can be used to further highlight different hierarchical levels. Thereby, the intensity of shadings is proportional to the depth of the hierarchy, i.e., the lower the level, the more intensive the shading will become. An illustration of this can be found below.

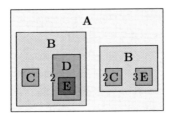

Since assigning attributes to species is a central feature of ML-Rules, the graphical syntax also allows for expressing the according aspects. However, as there is no such obviously good visual alternative like in the case of nested structures, the same notation like in the textual syntax is used here. This has also the advantage of keeping burdens low when switching between textual and graphical notations. Therefore, the syntax is as follows:

$$\mathbf{X}(A_{\mathbf{X}})[S_{\mathbf{X}}] \triangleq \boxed{\begin{array}{c} \mathbf{X}(A_{\mathbf{X}}) \\ S_{\mathbf{X}} \end{array}}$$

where $A_{\mathbf{X}}$ is the set of attribute values of species $\mathbf{X}$. The representation of attributed species within a nested hierarchy is straightforward:

$$\mathbf{B}(0.5, \mathbf{p}, 2)[\mathbf{C}(0) + 2\,\mathbf{D}[\mathbf{E}]] \triangleq$$

Please note, similar to the textual syntax, parentheses may be omitted if the arity $ar(\mathbf{X}) = 0$, i.e., if species of type $\mathbf{X}$ do not have assigned attributes. In the above example, this is the case for species $\mathbf{D}$ and $\mathbf{E}$, unlike species $\mathbf{C}$ and $\mathbf{B}$ with one and three attributes respectively.

The presented graphical notation of nested species is used for both, specifying the initial solution of a model as well as reactants and products of rules. Unlike various other visual modeling languages, interactions between different entities are not explicitly encoded by lines and arrows connecting the interaction partners; in Petri nets, for instance, interactions are visualized by arcs connecting place nodes with transitions. Instead, here reaction rules are still defined in terms of schematic and small modular units that may be applied to various contexts. Hence, the employed approach is similar to Bigraphical Reactive Systems in this point and therefore – apart from the specification of reactants and products – the notation of a graphical rule is similar to the previously introduced textual notation. This also implies that not only defined values, but also variables and expressions can be inserted in a similar way, as shown in the following example:

$$\boxed{\begin{array}{c} \mathbf{B}(x,\mathrm{p},2) \\ \boxed{\mathbf{C}(x/3)} \end{array}} \quad \xrightarrow[x\geq3]{x^2} \quad \boxed{\begin{array}{c} \mathbf{B}(0,\mathrm{u},2) \\ \boxed{\mathbf{C}(x-1)} \end{array}}$$

Compared to the description of initial solutions, specifying graphical rule schemata may require some few additional notations: Species identifiers for accessing the amounts of bound reactants are placed in the top-right corner of boxes and in case of binding remainder-solutions to variables, boxes with *dashed borders* are introduced to underline the semantical difference in comparison to regular boxes:

This graphical rule schema is equivalent to the textual representation of species identifiers and bound remainder-solutions as has been presented in Chapter 6, i.e.:

$$\mathbf{B}[\mathbf{C}^c + s?]^b \xrightarrow{k\#b\#c} \mathbf{B}[\mathbf{E} + s?] \ .$$

7.2 Names and Unspecified Attribute Sets

7.2.1 Attribute Names

So far, different attributes of a species are identified by their positions rather than names. This has the advantage of keeping notations compact, as merely expressions (like values and variables) need to be specified instead of expressions *and* names. However, when it comes to realistic applications, this method might have some drawbacks. Therefore, an alternative approach is presented here, where each species name is assigned a set of attribute names rather than just defining the number of attributes. For example, species $\mathbf{X}$ comprising $n \in \mathbb{N}$ attributes is no longer defined by its arity $ar(\mathbf{X}) = n$ but instead by $\mathbf{X}(x_0, \ldots, x_n)$, where each $x_i, i \in \{0, \ldots, n\}$ denotes a distinct attribute name of species $\mathbf{X}$. As part of the specification of the initial solution or within rules, attributes can then be referenced by $name^{expr}$, where *name* is the attribute's name and *expr* an expression that may be also a defined value or a simple variable, for instance.

In particular in case of having species with arities larger than one, identifying attributes by name may be less error-prone, since with defined names the order of specifying attributes does not matter. Models may also become more easy to understand if appropriate attribute names are chosen and thereby a model description consists of additional pieces of information. For example, without the need for further annotation it becomes obvious that the following species shall describe a certain protein A consisting of each a phosphorylation as well as methylation site of which the first one is in an unphosphorylated state and the second modification site is methylated:

$$\textbf{Protein\_A}(phosphorylated^{\texttt{false}}, methylated^{\texttt{true}}) \ .$$

By contrast, the nameless approach is clearly less verbose and informative:

$$\textbf{Protein\_A}(\texttt{false}, \texttt{true}) \ .$$

7.2.2 Don't Care, Don't Write

Besides its clear advantages, a drawback of identifying attributes by name is that this method leads to larger model descriptions, as can be already seen with this truly small example above. However, like it is done in BNGL, for instance, the existence of attribute names allows for a *"don't care, don't write"* approach, where attributes may be left unspecified if they are of no interest in certain situations. Therefore, analogous to binding remainder-solutions (see page 144f), we introduce the concept of binding *remainder-attributes* within rule schemata.

The idea is to bind arbitrarily many attributes to a special variable, so that certain attributes (i) do not need to be explicitly specified and (ii) the bound values can nonetheless be reused for assignments to product species on the rule's right-hand side. Alternatively, one could consider here also an implicit preservation of unspecified attributes without the need for binding them to a special variable (cf. a similar discussion on alternative choices for preserving sub-solutions on page 146). However, then in certain situations we would again loose full and transparent control about which defined attribute values shall be preserved, e.g., when a rule like

$$\mathbf{A}(x^{100}) + \mathbf{A}(x^{50}) \rightarrow \mathbf{A}(x^{150})$$

is applied to solution $S = \mathbf{A}(x^{50}, y^{\text{true}}) + \mathbf{A}(x^{100}, y^{\text{false}})$. It would be unclear whether attribute y of the resulting product species is assigned the value true or false. Therefore, we prefer to make the binding of sets of remainder-attributes explicit, similar to the concept of binding remainder-solutions.

The following example illustrates how the approach is intended to work. Let us assume a species $\mathbf{Cell}(xpos, ypos, volume, phase)$, whose attributes denote the two-dimensional spatial coordinates, the volume, and the phase of the division cycle of a cell. An increase of the volume independently from other attributes can then be described by either inserting an own variable for each attribute, for example,

$$\mathbf{Cell}(xpos^x, ypos^y, volume^v, phase^p) \rightarrow \mathbf{Cell}(xpos^x, ypos^y, volume^{v+\Delta}, phase^p),$$

or by leaving all attributes unspecified – except of the volume – and binding the set of remainder-attributes to a special variable, *remain?* in the example below:

$$\mathbf{Cell}(volume^v, remain?) \rightarrow \mathbf{Cell}(volume^{v+\Delta}, remain?).$$

Please note, specifying a kinetic rate and binding a possibly enclosed sub-solution has been omitted for simplicity reasons.

The *"don't care, don't write"* approach not only enables more succinct model descriptions, but makes it also more easy to extent models. For instance, if the above $\mathbf{Cell}$ species shall be later modified by additional attributes, e.g., to represent its spatial location within three dimensions rather than just two, the first rule schema would also require to be accordingly modified, while the second rule schema comprising of the variable *remain?* may remain the same.

7.2.3 Transformation to Basic Syntax

We have seen that introducing attribute names and binding arbitrarily large sets of unspecified attributes to a special variable facilitates modeling in various ways. The approach therefore denotes a meaningful extension of the basic ML-Rules language concept. Fortunately, like the graphical syntax presented in Section 7.1, the extension can be realized as purely syntactic sugar, i.e., a syntax can be defined that translates to the basic syntax of Chapter 6 and therefore does not require any change in the semantics of ML-Rules. The syntax and translation process, which may be straightforwardly integrated also into the graphical syntax, is defined as follows.

Specification of attribute names

Each attributed species name $\mathbf{X}$ is associated with an ordered set A_X of attribute names $x_i, i \in \{1, \ldots, n\}$, where $n \in \mathbb{N}$ is the total number of different attributes of species $\mathbf{X}$. Each attribute name in A_X is assumed to exists only once, i.e., there are no two names x_i and x_j for which $x_i = x_j$ holds true. However, different species names may have assigned same attribute names. We write $\mathbf{X}(x_1, \ldots, x_n)$ to specify the set of attribute names of species $\mathbf{X}$.

By contrast, identification of attributes by their position only requires a specification of the arity $ar(\mathbf{X})$ of species name $\mathbf{X}$. A translation to the basic syntax is therefore rather simple, as the number of different attribute names of $\mathbf{X}$, i.e., the number of elements of set A_X, translates to $ar(\mathbf{X})$:

$$\mathbf{X}(x_1, \ldots, x_n) \implies ar(\mathbf{X}) = n$$

From names to positions (and back)

Within rules and specifications of initial solutions, two important differences exist between attributed species (or species patterns) given in the extended syntax and the basic nameless approach. First, while in the basic syntax only a set of expressions e is given, the syntactical extension requires a set of tuples (x, e), written as x^e, where x denotes an attribute name. The second difference is that the order of elements matters in the basic syntax, while due to the identification by names in the extended syntax an unordered set of elements is given, i.e., the order of tuples (x, e) does not matter.

Therefore, the transformation process from the extended to the basic syntax eliminates attribute names and requires an ordering of the respective expressions. The order of the names of attributes is thereby defined by the previously specified set of attribute names of according species. That means, given the specification of species $\mathbf{X}(x_1, x_2)$ comprising two attributes with names x_1 and x_2, i.e., $A_X = \{x_1, x_2\}$, a translation to the nameless syntax always yields the same result, no matter in which defined order the attributes have been referenced by name:

$$\mathbf{X}(x_1^{e_1}, x_2^{e_2}) \implies \mathbf{X}(e_1, e_2)$$
$$\mathbf{X}(x_2^{e_2}, x_1^{e_1}) \implies \mathbf{X}(e_1, e_2)$$

We thereby assume again that each attribute name appears only once, i.e., $x_1 \neq x_2$, that each x_i is part of A_X, and that the number of specified attributes equals the number of elements of A_X. Otherwise, the model description would be invalid.

The presented transformation process needs to be done for each appearance of attributed species, i.e., reactants and products within rules as well as species being part of the initial solution. A transformation in the opposite direction – from positions to names – would be straightforwardly achieved by generating indexed attribute names x_i and tuples (x_i, e_i), where i denotes the position of an expression e. Therefore, the above example transformation could be inverted as follows:

$$\mathbf{X}(e_1, e_2) \implies \mathbf{X}(x_1^{e_1}, x_2^{e_2})$$

Determining unspecified attributes

Attributes of reactant and product species within rules may be ignored by explicitly referencing only a subset of attributes and binding the remainder – i.e., the set of unspecified attributes – to a special variable in ?-suffix notation. The translation of according patterns into the basic syntax therefore requires two major steps: First the unspecified sets of attributes need to be determined before subsequently a transformation from attribute names to positions can be performed in a way that has been presented above. Thereby, in the first step it is important to distinguish between reactant and product species.

To determine unspecified attributes of a reactant species $\mathbf{X}$, first of all the set A_S of *explicitly specified* attribute names is derived, e.g., the notation of reactant species $\mathbf{X}(x_1^{e_1}, x_2^{e_2}, remain?)$ with specified attribute names x_1 and x_2 leads to $A_S = \{x_1, x_2\}$. The next step is to compare the cardinality of set A_S

with the total number of attributes of species $\mathbf{X}$, i.e., $|A_X|$. Thereby, three distinct cases need to be considered:

$|A_S| > |A_X|$: Invalid description, process aborts

$|A_S| = |A_X|$: No unspecified attributes

$|A_S| < |A_X|$: A set of unspecified attributes needs to be determined

Only in the last case an unspecified set of attributes has been detected and thus needs to be determined. Here – due to the fixed arity – the notation of bound remainder-attributes is obligatory. The first case is the result from an invalid notation, since more attributes have been specified than species $\mathbf{X}$ consists of. If $|A_S|$ equals $|A_X|$, each attribute has been specified and thus neither a set of unspecified attributes needs to be determined nor the definition of a special variable for binding remainder-attributes is necessary. However, the latter is still allowed and might be useful in case of extending the model by additional attributes in the future, like it has been discussed on page 163. Again, we always assume that A_S is a subset of A_X or that $A_S = A_X$. Otherwise, the model specification is invalid.

If an unspecified set of attributes has been detected, we need to resolve it appropriately by first identifying the names of attributes that are part of it:

$$A_U = A_X \backslash A_S \ .$$

Then, for all $x \in A_U$ a tuple (x, v) – written as x^v – is added to a set Rem, where x is an unspecified attribute name and v a variable for binding the attribute's value. Finally, the special variable that binds the set of unspecified remainder-attributes is replaced by Rem and a transformation from names to positions is applied to the resulting reactant species. For instance, given the specification of species $\mathbf{X}(x_1, x_2, x_3)$, three example translations of different reactant patterns (including intermediate steps) are shown below.

$$\mathbf{X}(x_1^{e_1}, x_2^{e_2}, remain?) \implies \mathbf{X}(x_1^{e_1}, x_2^{e_2}, x_3^{v_3}) \implies \mathbf{X}(e_1, e_2, v_3)$$
$$\mathbf{X}(x_2^{e_2}, remain?) \implies \mathbf{X}(x_2^{e_2}, x_1^{v_1}, x_3^{v_3}) \implies \mathbf{X}(v_1, e_2, v_3)$$
$$\mathbf{X}(remain?) \implies \mathbf{X}(x_1^{v_1}, x_2^{v_2}, x_3^{v_3}) \implies \mathbf{X}(v_1, v_2, v_3)$$

The transformation of product species is simpler compared to reactants, as here sets of unspecified attributes rely on previously determined sets of reactants. Therefore, a variable of bound remainder-attributes simply needs to be replaced by the according set Rem and afterwards we only need to check for consistency, such that $A_S \cup A_U = A_X$ holds true, where A_U is the set of unspecified attributes determined from one of the reactant species of the same rule. Given the same specification of species $\mathbf{X}$ as in the previous example, the transformation of an entire rule schema – in which only attribute x_2 is of interest – may thus look as follows:

$$\mathbf{X}(x_2^e, remain?) \rightarrow \mathbf{X}(x_2^{e'}, remain?) \qquad \Longrightarrow \qquad \mathbf{X}(v_1, e, v_3) \rightarrow \mathbf{X}(v_1, e', v_3) \,.$$

7.3 Generic Species

The hierarchy-based pattern matching of ML-Rules (see Chapter 6.2.8, beginning at page 142) allows a single rule schema to be applied to various contexts, e.g., to concisely describe reactions that may take place within different compartments. If those reactions depend on contextual information, however, the surrounding "compartment" species needs to be explicitly specified, which counteracts the goal of achieving succinct and easily extensible models. Therefore, a *generic* or *wildcard* species name may be useful in such situations, especially in combination with unspecified attributes.

Let us assume, for instance, a bimolecular reaction "A+B $\rightarrow$ C" taking place within different compartments, e.g., the cytoplasm and the endosome. To correctly describe the volume-dependent kinetics of this reaction (cf. page 51f), we need two distinct rules that either comprise different reaction rate coefficients (only applicable for describing compartments with constant volumes) or that are taking certain attributes of the compartment species into account and thereby also allow for dynamic volumes:

$$\mathbf{Cyto}(vol^v)[\mathbf{A}^a + \mathbf{B}^b + r_S?] \xrightarrow{\frac{k}{v}\#a\#b} \mathbf{Cyto}(vol^v)[\mathbf{C} + r_S?]$$

$$\mathbf{Endo}(vol^v)[\mathbf{A}^a + \mathbf{B}^b + r_S?] \xrightarrow{\frac{k}{v}\#a\#b} \mathbf{Endo}(vol^v)[\mathbf{C} + r_S?] \,.$$

To avoid writing two nearly identical rules, we could replace both species $\mathbf{Cyto}(vol)$ and $\mathbf{Endo}(vol)$ by a species $\mathbf{Comp}(name, vol)$, whose *name* attribute denotes whether $\mathbf{Comp}$ represents the cytoplasm or an endosome compartment. In this case, a single rule schema like the following example would suffice:

$$\mathbf{Comp}(name^n, vol^v)[\mathbf{A}^a + \mathbf{B}^b + r_S?] \xrightarrow{\frac{k}{v}\#a\#b} \mathbf{Comp}(name^n, vol^v)[\mathbf{C} + r_S?] \ .$$

On the other hand, this does not denote a suitable approach if species **Cyto** and **Endo** are possibly part of other rules as well, as then all these rules would also require according modifications. Moreover, different compartments might comprise different attributes, e.g., in addition to a volume, the endosome compartment might also be characterized by a dynamically changing pH value. That means, one universal set of attributes might be unsuitable to describe diverse properties of different compartments. Therefore, another approach is presented here that allows for specifying a generic species name. The idea is to use wildcards instead of defined species names in order to reduce the number of rules needed. Thereby, according to other language constructs in ML-Rules which also encode for unspecified aspects, again a notation applying the "?" symbol is chosen:

$$?(vol^v, r_A?)[\mathbf{A}^a + \mathbf{B}^b + r_S?]^{comp} \xrightarrow{\frac{k}{v}\#a\#b} comp(vol^v, r_A?)[\mathbf{C} + r_S?] \ .$$

The generic reactant $?(vol^v, r_A?)[\dots]$ matches every species that has assigned an attribute vol, but in addition a set of unspecified attributes may exist. Therefore, according species may comprise different sets of attribute names, such as **Cyto**(vol) and **Endo**(vol, pH). Please notice, the assigned identifier $comp$ is used to determine the name of the product species. By doing so, defined relations between reactants and products can be specified, which might be of particular importance if a rule schema consists of more than one generic species. The idea of generic species shares many similarities with the basic principles of classification and subsumption from *description logic* (Baader et al., 2003) to determine whether a certain object or concept is member of a class of objects or concepts.

Generic species are not only useful for modeling dynamic processes with contextual information, as has been exemplified above, but also the description of reactions at a single level may profit from this approach. For example, different proteins often consist of a conserved binding motif, i.e., a binding site that is highly similar among different molecular species. Thereby, different proteins are able to bind the same ligand. An example for a highly conserved and well-known binding motif is the guanosine triphosphate (GTP) binding motif of a family of proteins called G proteins (Hamm, 1998). Instead of explicitly specifying numerous of nearly identical reaction rules, of which each describes the binding of GTP to one specific kind of G protein, a single rule schema comprising a generic species may encode for the same dynamics, e.g.,

$$\mathbf{GTP}^g + ?(gtp^{\mathtt{false}}, r_A?)^{gprot} \xrightarrow{k\#g\#gprot} gprot(gtp^{\mathtt{true}}, r_A?) \ .$$

To enrich ML-Rules with the presented functionality, again a transformation to an alternative syntax denotes a viable option, in this case a transformation to the extended syntax allowing for names and unspecified sets of attributes presented in Chapter 7.2. Therefore, a rule comprising generic species translates to a *set of rules*, where the wildcard name on the left-hand side (i.e., the "?" symbol) as well as reused identifiers on the product side are replaced by defined species names. Each generated output rule thereby comprises different replacements, and if attributes are specified, only those names from the set of all species names are considered that match the according set of explicitly specified attributes. That means, given the following specification of species names and attributes:

$$\mathbf{A}(\,), \ \mathbf{B}(\,), \ \mathbf{C}(\,), \ \mathbf{Cyto}(vol), \ \mathbf{Endo}(vol, pH),$$

the transformation process of a generic rule schema

$$?(vol^v, r_A?)[\mathbf{A}^a + \mathbf{B}^b + r_S?]^{comp} \xrightarrow{\frac{k}{v}\#a\#b\#comp} comp(vol^v, r_A?)[\mathbf{C} + r_S?]$$

will replace the wildcard by names **Cyto** and **Endo** only. **A**, **B**, and **C** are not considered, as these species lack an attribute vol. The transformation process therefore yields the following set of rule schemata:

$$\mathbf{Cyto}(vol^v, r_A?)[\mathbf{A}^a + \mathbf{B}^b + r_S?]^{comp} \xrightarrow{\frac{k}{v}\#a\#b\#comp} \mathbf{Cyto}(vol^v, r_A?)[\mathbf{C} + r_S?]$$

$$\mathbf{Endo}(vol^v, r_A?)[\mathbf{A}^a + \mathbf{B}^b + r_S?]^{comp} \xrightarrow{\frac{k}{v}\#a\#b\#comp} \mathbf{Endo}(vol^v, r_A?)[\mathbf{C} + r_S?]$$

By contrast, a generic rule schema like $?(r_A?)^x \xrightarrow{k\#x} \emptyset$ encodes for a larger set of rules comprising each defined species name:

$$\mathbf{A}(r_A?)^x \xrightarrow{k\#x} \emptyset$$

$$\mathbf{B}(r_A?)^x \xrightarrow{k\#x} \emptyset$$

$$\mathbf{C}(r_A?)^x \xrightarrow{k\#x} \emptyset$$

$$\mathbf{Cyto}(r_A?)^x \xrightarrow{k\#x} \emptyset$$

$$\mathbf{Endo}(r_A?)^x \xrightarrow{k\#x} \emptyset$$

This shows that the presented approach of generic species names is a powerful tool, which should be handled with care – preferably in combination with appropriate attributes – to narrow down its application.

7.4 Functions on Solutions

Whereas the language concept of ML-Rules so far allows the use of arbitrary functions on attributes, the application of *functions on solutions* is not yet supported. However, since such functions may facilitate the description of multilevel dynamics (Maus et al., 2011), a discussion on relevant aspects and needs is provided in the following. It is also examined how this feature could be integrated into ML-Rules.

7.4.1 Problem Statement

Multilevel modeling requires that states at higher and lower levels are observable in order to take effects for the behavior at certain other levels of organization. As has been explained before, such phenomena are called downward and upward causation (see Sections 2.1.3 and 3.3, for instance).

In the case of downward causation, an observation of states at higher levels can simply be achieved by expanding the context (or scope) of a rule toward environmental species. Thereby, attributes representing high-level states become accessible and can thus be taken into account for describing dynamic behavior at lower levels. In the preceding Chapter 7.3, for example, the volume of a cellular compartment determines the rate of an enclosed bimolecular reaction. States at lower levels, by contrast, are typically not represented by distinct attributes but are the result of collective behavior of many small parts, i.e., the sum of individual states of species at a lower level. The question is, how can we retrieve the according information in order to describe upward causation?

A typical example for upward causation is the amount of a particular species influencing behavior at a higher level of organization. Using *species identifiers* in combination with the *number sign* operator ($\#$) is a simple means for accessing the amounts of species in ML-Rules, which is – by the way – not only important for modeling upward causation but also for generally specifying dynamic rate kinetics in ML-Rules, such as mass-action kinetics of elementary biochemical reactions. However, this method requires each species – whose amount we are interested in – to be defined as a reactant of the rule, no matter whether it will be consumed or modified when the rule fires. Consequently, the rule is not allowed to fire in the absence of any of these reactants, as then the firing condition is unfulfilled. In some situations this can be problematic, e.g., for the specification of certain non-mass-action kinetics (see Chapter 8, page 182). Also, by using species identifiers, it is difficult to aggregate the amounts of different incarnations of attributed species, e.g., if we would be interested in the *total* amount of a certain protein rather than in the amounts of differently phosphorylated states of it. The reason for that lies in the instantiation of rules, as, e.g., a reactant pattern $\mathbf{Prot}(x)^p$ applied to a solution

$$S = 5\,\mathbf{Prot}(\text{phos}) + 2\,\mathbf{Prot}(\text{unphos})$$

yields two different rule instantiations with different kinds of species $\mathbf{Prot}(x)$ bound and therefore – depending on the concrete instantiation – the expression $\#p$ evaluates to values 5 or 2 rather than to the aggregated total amount of 7.

So, in some situations species identifiers are not suitable for retrieving certain information from a solution. To overcome this limitation, we could adopt ideas of the imperative π-calculus or BNGL and introduce the concept of special global observables keeping track of the amounts of certain species. However, as has been shown on pages 81 and 86f in Chapter 4.2, using *global* variables for modeling multilevel systems may cause problems with respect to combinatorial complexity and modeling dynamic structures. Hence, developing a method for integrating some kind of *contextual* or *private* observables would be required, whose visibility is restricted to a given context. On the other hand, ML-Rules already allows for binding sub-solutions that are enclosed by defined species, i.e., in a context-dependent manner. Thus, extending the language by specific functions for counting species in a given solution seems to denote a more straightforward alternative (see Section 7.4.2).

Besides the need for retrieving information, functions on solutions may be also of importance for describing certain variable structure dynamics. So far, bound solutions can be freely reused on the product-side of a rule, which facilitates the description of phenomena like migration, merging, and copying of large sets of species. However, changing the composition of solutions flexibly is another matter, for which additional operations would be required, e.g., to split a solution equally into two new solutions or to remove or modify the entire population of a certain species within a solution while the remainder stays untouched. Such operations may be important to describe processes like cell division, budding, and simultaneous behavior of a collective of species, for instance.

Other functions might be useful for modeling rather specific systems. For example, encoding an individual-based RNA structure folding model, where each nucleotide of an RNA strand is represented by a distinct molecular species and the dynamic process of RNA folding is described by continuing base-pair closing and reopening events among individual nucleotides, requires macro knowledge about the current structure in order to determine which nucleotides may pair and which not (Maus, 2008). A function that provides this information by iterating over the elements of the RNA solution would clearly facilitate this endeavor. However, supporting such model-specific functions would require an integration of user-defined function libraries. Therefore, in the following only more generally applicable functions are discussed.

7.4.2 Counting Function

The need for flexibly counting the elements of a solution has been identified and accordingly addressed by others as well. In the React(C) language, for instance, arbitrary functions on solutions can be defined with the help of λ-

calculus terms and some additional terms for handling molecules and solutions (John et al., 2011). This allows, e.g., to define a function that receives an attributed molecule and a solution to count the occurrence of this molecule within the respective solution. This general idea of counting is adopted here by extending ML-Rules by an according parameterizable function.

Such a special function for counting certain species within a given solution is relatively easy to integrate into the general language concept of ML-Rules. This is due to the fact that applying the function does *not change* the solution but only *retrieves* information from it and can thus be treated and used like other functions within mathematical expressions.

The parameterizable counting function takes two arguments: a species pattern *pat* as well as a solution *sol* (the latter must not be explicitly specified by the user; hence, it is typically passed in the form of a bound variable). An evaluation of the function always yields a non-negative integer value. The function's notation is defined as

$$\sum pat \in sol$$

where the syntax of *pat* is the same like that for specifying reactant patterns.

By iterating over the elements of *sol*, this function aggregates the amounts of all species matching the specified pattern. Thereby, however, in contrast to the process of rule instantiation, no *deep* pattern matching is done, i.e., although the passed solution may consist of a nested hierarchy, here the species pattern is compared with the topmost level only (cf. Section 6.2.8). For example, let us assume a solution

$$sol? \ = \ 3\,\mathbf{A}(0) \ + \ 10\,\mathbf{A}(1) \ + \ 2\,\mathbf{A}(0)[6\,\mathbf{B}] \ + \ \mathbf{C}[3\,\mathbf{A}(0)] \ .$$

In this case, an expression $\sum \mathbf{A}(x) \in sol?$ evaluates to a species count of 15, since the pattern $\mathbf{A}(x)$ does not specify a defined attribute value and thus each incarnation of species $\mathbf{A}$ is matched, if found at the topmost level; not considered is the amount of species $\mathbf{A}(0)$ that is enclosed by $\mathbf{C}$.

Expression $\sum \mathbf{A}(1) \in sol?$, by contrast, which counts all species $\mathbf{A}$ that have assigned an attribute 1, evaluates to a species count of 10. Please notice, the pattern *pat* may also describe a nested species. Accordingly, the evaluation of expression $\sum \mathbf{A}(x)[\mathbf{B}] \in sol?$ returns the value 2.

7.4.3   Splitting of Solutions

To split or decompose a given solution into smaller ones, a special function "split(sol, pct)" could be introduced, which takes a solution sol and a percentage pct as arguments and returns a new solution consisting of $pct\%$ of the elements of sol. This way, we could describe a cell division process, where both daughter cells consist of half of the original solution:

$$\textbf{Cell}[sol?] \rightarrow \textbf{Cell}[\text{split}(sol?, 50)] + \textbf{Cell}[\text{split}(sol?, 50)]$$

However, in this case, undesired effects may appear due to multiple independent invocations of the function, e.g., species with low copy numbers in the solution may easily vanish. The content of a dividing cell in reality, by contrast, becomes completely partitioned, such that the sum of the content of both daughter cells equals the original content. Therefore, although requiring slight adaptations of the syntax and semantics of ML-Rules, a flexible decomposition method where nothing gets implicitly lost seems to be a better approach.

Universal decomposition approach

By using the "+" symbol, product solutions in ML-Rules can be arbitrarily composed from smaller sub-solutions. Thereby, it is also possible to freely combine explicitly specified parts (defined species) with arbitrarily many unspecified parts (bound remainder-solutions). The idea is to use a similar notation for the decomposition of solutions on the left-hand side of a rule.

To some degree, decomposition of solutions is already supported in ML-Rules, however, the approach so far is limited to one unspecified part at most, i.e., the entire unspecified remainder-solution, besides possibly multiple explicit specifications of defined reactants. This method is now extended in a way that allows for decomposing the remainder-solution by specifying multiple "?-variables" on the rule's left-hand side and thereby binding multiple unspecified sub-solutions. For example, the following rule

$$\textbf{Cell}[sol_1? + sol_2?] \rightarrow \textbf{Cell}[sol_1?] + \textbf{Cell}[sol_2?]$$

describes the splitting of a **Cell** species including its enclosed solution. Thereby, the content is split equally and both parts are bound to two distinct variables

sol_1? and sol_2?, which are used to assign each product **Cell** with one half of the original solution. Similarly, a rule's left-hand side "**Cell**[sol_1? + sol_2? + sol_3?]" would split the content of **Cell** equally into three distinct sub-solutions.

Splitting *equally* means to partition a solution with the intention that each resulting sub-solution consists of the same multiset of species, each of which with an amount of n/m, where n is the species' amount in the original solution and m the number of sub-solutions into which the solution is decomposed. Please notice, it might be possible that the resulting sub-solutions are not totally equal. This is always the case if – for at least one species – the Euclidean division of n by m leaves a remainder. If it is not possible to split the solution into entirely equal sub-solutions, the surplus will be randomly distributed among the set of sub-solutions, but the total amounts of top-level species among all sub-solutions will become as equal as possible.

However, in any case, the union of all sub-solutions equals the original solution, i.e., the sum of the partial amounts of any species equals the amount of the according species in the original solution. For example, if we apply the above rule of a dividing cell to a reactant species "**Cell**[12 **A**+3 **A**[**B**]+**C**[7 **A**]]", its enclosed solution may decompose in two different ways:

$$sol_1? = 6\,\mathbf{A} + 2\,\mathbf{A}[\mathbf{B}] \qquad\qquad sol_1? = 6\,\mathbf{A} + \mathbf{A}[\mathbf{B}] + \mathbf{C}[7\,\mathbf{A}]$$
$$\text{or}$$
$$sol_2? = 6\,\mathbf{A} + \mathbf{A}[\mathbf{B}] + \mathbf{C}[7\,\mathbf{A}] \qquad\qquad sol_2? = 6\,\mathbf{A} + 2\,\mathbf{A}[\mathbf{B}]$$

Please note, although different concrete decompositions may be possible, it does not yield different instantiations of the rule. That means, semantically, first the specified reactants of a rule are matched like it is explained in Chapter 6.2. Depending on this matching process, different rule instantiations may be generated, each of which has bound implicitly an according remainder-solution. Thereafter, this unspecified remainder will be decomposed and bound to according variables, so that the resulting sub-solutions can be used to specify the rule's product species, for instance. The random split of remaining parts may thereby denote an additional source for stochastic variability.

Directed decomposition

Above, a universal decomposition approach has be presented that splits a solution into arbitrarily many equal parts. However, sometimes it might be desired to direct or control the decomposition, such that, e.g., one resulting sub-solution is larger than another one or a particular sub-solution consists of species of a certain kind of species only. Therefore, an additional operator "◁" is introduced.

The fraction size of different parts of a decomposed solution can be controlled by an optional term "◁ f" with $f \in \mathbb{R} \mid 0 \leq f \leq 1$ behind the names of variables binding the according sub-solutions. Thereby, the sum of all fractions f must be 1. For example, the following rule splits the content of a **Cell** species into two sub-solutions, of which "sol_2?" contains only 10% of the total amounts of species enclosed by the **Cell**:

$$\mathbf{Cell}[sol_1? \triangleleft 0.9 + sol_2? \triangleleft 0.1] \rightarrow \mathbf{Cell}[sol_1?] + \mathbf{Cell}[sol_2?] \ .$$

Such inhomogeneous splitting may be of importance to describe a budding process, for instance, where a cell is reproduced by a small outgrowth on its surface rather than a centered division.

Besides controlling the fraction size of sub-solutions, the "◁" operator can be also used to constrain the composition of a sub-solution to a particular species pattern. A combined control of species patterns and fraction sizes within the same rule is thereby not allowed. Also, at least one decomposition part must be unconstrained, to ensure that the union of all sub-solutions equals the original solution. For example, the following reactant pattern splits the content of a **Cell** species into three sub-solutions, of which one comprises all species **A** and the remainder is split equally into two other sub-solutions "sol_1?" and "sol_2?":

$$\mathbf{Cell}[sol_A? \triangleleft \mathbf{A}(x) \ + \ sol_1? \ + \ sol_2?] \rightarrow \ldots$$

More specific as well as nested species patterns are also possible. For example, "$sol_A? \triangleleft \mathbf{A}(1)[\mathbf{B}]$" yields a solution "$sol_A$?" consisting only of those species $\mathbf{A}(1)$ that enclose at least one species **B**.

Chapter 8

Case Studies

After introducing the multilevel modeling concept of ML-Rules, now it is time to see how well this rule-based language applies to describing realistic models of biological multilevel systems. Therefore, in Chapter 8.1, the Notch signaling example from Chapter 3 will be revisited at first. After that, an example of the fission yeast cell cycle regulation and mating type switching at multiple organizational levels will be given. Although both case studies rely on models developed for demonstration purposes rather than providing new biological findings, they can be nevertheless considered as exemplary representatives of biological multilevel models comprising of upward and downward causation within hierarchically nested multi-compartment structures.

The models in this chapter are generally specified in the extended syntax presented in Chapter 7, i.e., including attribute names and mostly in a graphical manner. However, the according basic textual notations are also given (if not in this chapter, they can be found in Appendix B). Wherever functions on solutions are used, possible alternative specifications excluding this language extension will be additionally discussed.

8.1 Notch Signaling Example

8.1.1 General Structure of the Model

As has been described in detail in Chapter 3, the Notch signaling model consists of multiple interacting cells, each of which has its own state and its own com-

Table 8.1: Species definitions of the ML-Rules Notch signaling model.

Biological structure / entity	Species name	Attribute names
Entire cell	**Cell**	$(id, phase)$
Membrane	**Mem**	(vol)
Cytoplasm	**Cyt**	(vol)
Nucleus	**Nuc**	$(\,)$
Notch receptor	**N**	$(state)$
Delta ligand	**D**	$(state)$
Free space	**FS**	(id)

partmentalized intracellular dynamics of Notch receptors and Delta ligands. Therefore, besides "molecular" species representing Notch and Delta proteins, also different "compartment" species are defined to appropriately reflect the crucial structural elements of the system (Table 8.1). Both the membrane (**Mem**) and cytoplasm (**Cyt**) species thereby consist of a volume attribute vol, while the nucleus species (**Nuc**) does not have attributes. **Cell** species are characterized by two different attributes representing a cell's discrete one-dimensional spatial coordinate (id) and its cell cycle $phase$ respectively. Delta (**D**) and Notch (**N**) species both have a $state$ attribute that is intended to represent ligand-receptor binding or an unbound state. In the case of unbound Notch, the $state$ attribute additionally denotes whether the molecule is a mature receptor molecule consisting of both the intracellular as well as extracellular domain or whether it has been cleaved and therefore consists of merely the intracellular domain (Nicd). Finally, a species **FS**(id) is defined to represent "free" spatial locations. Its usage for the process of cell division will be explained in Section 8.1.3.

The model may be differently initialized, however, a valid initial solution, i.e., the model's initial state, is shown in Figure 8.1. It consists of several individual **FS** species and at least one **Cell** species. The id attributes of **FS** and **Cell** species are thereby assumed to be assigned with unique integer values, so that each concrete species appears only once. The initial $phase$ of the cell has assigned the value G1. In addition, the **Cell** species encloses one

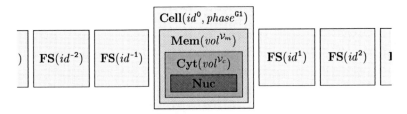

Figure 8.1: Initial solution of the ML-Rules Notch signaling model.

species $\mathbf{Mem}(vol^{\mathcal{V}_m})$, which consists of a species $\mathbf{Cyt}(vol^{\mathcal{V}_c})$, which in turn consists of a species $\mathbf{Nuc}$. Notch and Delta species are not part of this initial solution. They are dynamically produced by according rules.

8.1.2 Biochemical Processes

Most dynamics of the Notch signaling model can be classified as biochemical processes. To those belong classical elementary reactions, abstract processes like gene expression, and the translocation of molecules from one compartment into another. These processes are described in terms of rule schemata, so that each rule may be instantiated in various contexts such as different cells or different compartments with varying size and content. The set of rules correlates more or less with the list of intercellular and intracellular biochemical reaction equations from page 45 (Table 3.1). However, the rules here account for the explicitly nested model structure and causalities between different levels.

Let us begin with the translocation of Delta and Notch proteins from the cytoplasmic compartment to the membrane, which is described as follows:

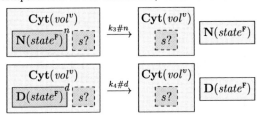

The attribute value F thereby represents a free receptor or ligand molecule respectively. We could additionally draw a box $\boxed{\mathbf{Mem}(vol^{\mathcal{V}_m})}$ around the reactant and product solutions to specify the membrane compartment explicitly.

However, this is not needed here, since the hierarchy of compartment species within each cell is assumed to be fixed and thus is at any time as has been initially defined (see Figure 8.1). Please note, as the **Cyt** compartment species within each cell – like the other compartments – is assumed to exist only once, there is no need for binding a species identifier and taking its amount into account within the rate expressions. As both above translocation rules look rather similar, it would be also possible to reduce the number of rules to a single generic rule schema, whose actual rate depends on the bound species p:

$$\mathbf{Cyt}(vol^v) \quad \boxed{?(state^F)}^p \quad \boxed{s?} \xrightarrow{\text{if } p=\mathbf{N}(state^F) \text{ then } k_3\#p \text{ else } k_4\#p} \quad \mathbf{Cyt}(vol^v) \quad \boxed{s?} \quad \boxed{p(state^F)}$$

The next step is to specify rules for receptor-ligand binding and subsequent cleavage of activated Notch. Since binding of Delta ligands by Notch receptors is restricted to membrane-residing molecules of adjacent cells, the according rule schema consists of species with an extended context, i.e., the surrounding species **Mem** and **Cell** are also part of the rule. This way, the id attributes of two **Cell** species can be compared to constrain the binding reaction to Notch and Delta within neighboring cells. In addition, the volume-dependent rate of this bimolecular reaction needs to take the vol attributes of both membranes into account. On the right-hand side of the rule, values F representing unbound proteins are replaced by variables j and i respectively. Thereby, $\mathbf{N}(state^j)$ within $\mathbf{Cell}(id^i, a_j?)$ is assigned the id of the other (adjacent) cell and vice versa:

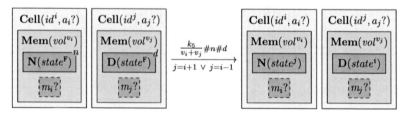

Please notice, the above rule assumes that nothing but a nested **Mem** species is enclosed by species **Cell**. Otherwise, preserving the unspecified content of both cells would require a binding of according remainder-solutions.

The subsequent cleavage of activated receptors, i.e., of Notch molecules that bind a Delta ligand, again requires an extended context comprising the according **Cell** species. In addition, the cytoplasmic compartment (**Cyt**) of the Notch-containing cell is explicitly specified, as this is the location of the intracellular domain (Nicd) of the cleaved Notch receptor represented by $N(state^I)$. Accordingly, the Delta ligand has still bound the Notch extracellular domain (Necd), which is represented by species $D(state^E)$:

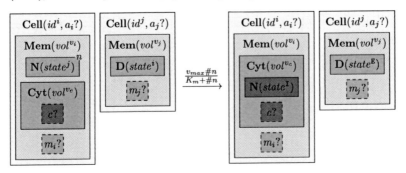

Please notice the rule's rate, which reflects the catalytic process by following Michaelis-Menten kinetics.

Both products of the previous rule may subsequently migrate to a lower level of the composition hierarchy, i.e., cytoplasmic Nicd may translocate into the nucleus and the membrane-anchored Delta ligand that is bound to Necd may be recycled, which means that the ligand is engulfed into the cytoplasm and Necd is removed. The respective rules are as follows:

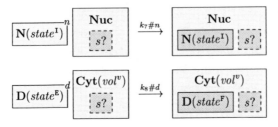

As the rate coefficient for protein degradation (k_{deg}) is assumed to be the same, no matter which protein is degraded and within which compartment, a single rule schema comprising a generic reactant species is sufficient to describe degradation of unbound Delta and Notch in different states and within different locations:

$$\boxed{?(state^x)}^{\!\!\!-p} \xrightarrow[x \in \{\text{F,I,E}\}]{k_{deg}\#p} \emptyset$$

The intercellular signaling process is completed with the synthesis of Notch and Delta described by abstract gene expression processes. As has been introduced in Chapter 3.1.3, both proteins are therefore assumed to be synthesized in dependence on the nuclear amount of Nicd and inserted into the cytoplasmic compartment:

$$\boxed{\begin{array}{c}\textbf{Nuc}\\ \boxed{s?}\end{array}} \xrightarrow{k_1'+k_1\theta} \boxed{\begin{array}{c}\textbf{Nuc}\\ \boxed{s?}\end{array}}\boxed{\mathbf{N}(state^\text{F})}$$

$$\boxed{\begin{array}{c}\textbf{Nuc}\\ \boxed{s?}\end{array}} \xrightarrow{k_2(1-\theta)} \boxed{\begin{array}{c}\textbf{Nuc}\\ \boxed{s?}\end{array}}\boxed{\mathbf{D}(state^\text{F})}$$

with $\theta = \frac{[\text{Nicd}]^h}{K^h+[\text{Nicd}]^h}$ and $[\text{Nicd}] = \sum \mathbf{N}(state^\text{I}) \in s?$.

To avoid an application of the counting function on the bound solution ($s?$) for determining the amount of nuclear Nicd, the easiest way would be to specify $\mathbf{N}(state^\text{I})$ as a reactant within **Nuc**, which allows to access its current amount via species identifier n, like it is done in the following rule, for instance:

$$\boxed{\begin{array}{c}\textbf{Nuc}\\ \boxed{\mathbf{N}(state^\text{I})}^{\!-n}\boxed{s?}\end{array}} \xrightarrow{k_2\left(1-\frac{\#n^h}{K^h+\#n^h}\right)} \boxed{\begin{array}{c}\textbf{Nuc}\\ \boxed{\mathbf{N}(state^\text{I})}\boxed{s?}\end{array}}\boxed{\mathbf{D}(state^\text{F})}$$

However, this naive approach does not work here, since if the amount of $\mathbf{N}(state^\text{I})$ equals zero, the reactant pattern does not match and therefore the rule would not be allowed to fire, although nuclear Nicd denotes merely a side-condition and its absence does not affect the validity of the rule. Therefore, as an alternative, species **Nuc** is assigned an attribute *nicd* that holds the current amount of nuclear Nicd:

$$\boxed{\begin{array}{c}\textbf{Nuc}(nicd^n)\\ \boxed{s?}\end{array}} \xrightarrow{k_2\left(1-\frac{n^h}{K^h+n^h}\right)} \boxed{\begin{array}{c}\textbf{Nuc}(nicd^n)\\ \boxed{s?}\end{array}}\boxed{\mathbf{D}(state^\text{F})}$$

Please note, in this case, the rule for Nicd translocation into the nucleus needs to be adapted accordingly and also an additional rule for the degradation of nuclear Nicd is needed.

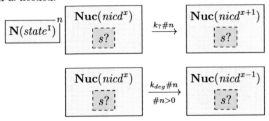

Hence, the above example emphasizes the importance of counting species within solutions.

8.1.3 High-Level Processes

Beyond various biochemical processes, the Notch signaling example model in this thesis includes three dynamic processes at higher organizational levels: The volume increase of membrane and cytoplasm compartments, a G_1 checkpoint determining whether the cell cycle proceeds with phase G_2 or will be arrested in phase G_0, and finally the process of cell division, which leads to an increasing number of cells in the model.

According to the informal description in Chapter 3.2, the cell is assumed to grow exponentially in size; hence the volumes of different cellular compartments are increasing over time. However, compared to the cytoplasm, the membrane compartment is assumed to increase only half as fast and the nuclear volume is assumed to remain constant. The growth of both compartment volumes can be described best by a single rule schema, since both processes rely on the same dynamics including a constraint restricting volumes to be doubled at most (twice the initial cytoplasmic volume $\mathcal{V}_c$):

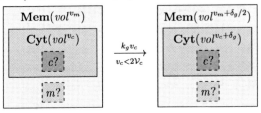

The next high-level rule is also constrained by the cytoplasmic compartment volume, as in our example model the G_1 checkpoint is assumed to not take place before the volume has increased by at least 50%:

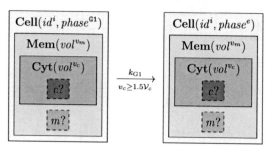

with expression $e :=$ if $(\sum \mathbf{N}(state^F) \in c?) < t_N$ then G2 else G0.

A conditional expression on the right-hand side of the rule checks whether the cytoplasmic amount of mature Notch receptors is lower or greater than a certain value t_N. If the amount equals or exceeds the threshold t_N, the cell's *phase* attribute is assigned the value G0, otherwise value G2 is assigned. An alternative rule without using the capability of counting species in a solution is given below:

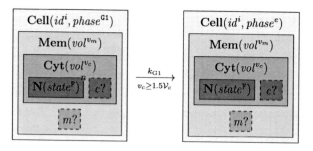

with expression $e :=$ if $\#n < t_N$ then G2 else G0.

Please note, with the above rule the aforementioned problem of zero-amounts might arise again, i.e., in the absence of cytoplasmic Notch the rule is not allowed to fire, although merely denoting a side-condition. However, since Notch is continuously synthesized and the rate k_{G1} is assumed to be rather high, the described situation should rarely occur and thus the problem can be neglected in this case. However, alternatively it would still be possible to equip the **Cyt** species with an additional attribute to store the amount of enclosed Notch, similar to what has been done with the **Nuc** species in order to describe gene expression correctly.

The final rule is dedicated to the process of cell division, which leads to the dynamic instantiation of additional cells in the model and thus denotes an illustrative example for the need of supporting variable model structures. Cells in the example model are considered to be linearly arranged in a one-dimensional space. Therefore, to represent different location and for modeling spatial dynamics like receptor-ligand binding between adjacent cells, species **Cell** is equipped with an *id* attribute. Each *id* is assumed to be an integer value appearing only once, by which the model's space becomes discretized. Hence, for modeling the process of cell division and thereby introducing a new cell instance, we need to determine whether a certain cell has already a direct neighbor or whether the space next to its own location is still "free". Therefore, the model is initialized with numerous **FS**(id) species representing different locations of free space (cf. Figure 8.1) and the process of cell division then requires an **FS** species whose *id* is adjacent to that of the dividing cell. If this is the case and all other conditions are also fulfilled, e.g., the dividing cell needs to be in phase G$_2$, reactant **FS** will be replaced by a newly created **Cell** species. The process of cell division is therefore generally described as follows:

$$\boxed{\begin{array}{c} \mathbf{Cell}(id^i, phase^{G2}) \\ \dots \end{array}} \quad \boxed{\mathbf{FS}(id^j)} \quad \xrightarrow[j=i+1 \,\vee\, j=i-1]{k_{G2}} \quad \boxed{\begin{array}{c} \mathbf{Cell}(id^i, phase^{G1}) \\ \dots \end{array}} \quad \boxed{\begin{array}{c} \mathbf{Cell}(id^j, phase^{G1}) \\ \dots \end{array}}$$

However, since cell division is assumed to take place after the cytoplasmic compartment volume has doubled, we need to explicitly include the according subcellular species **Cyt** in order to constrain firing based on its volume attribute *vol*. A specification of both the **Mem** and the **Cyt** species of a dividing cell is also required, as compartment volumes are assumed to be reset to their initial values $\mathcal{V}_m$ and $\mathcal{V}_c$ respectively. However, the question is how to deal with the enclosed sub-solutions? In the following, two alternative methods are presented.

A simple method would be to instantiate the "new" cell like it is done in the initial solution, i.e., without containing any Notch nor Delta. The dividing cell persists as it is, except of its cell cycle phase and compartment volumes that are reset to initial values:

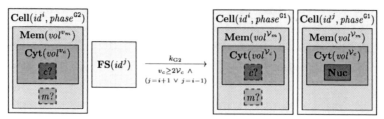

Alternatively, to split the content of a dividing cell equally and distribute it among both daughter cells, the universal decomposition approach presented in Chapter 7.4.3 can be used:

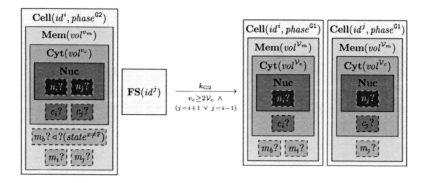

Therefore, multiple special variables for binding unspecified content are defined within each compartment, e.g., n_i? and n_j? within compartment species **Nuc**. By doing so, the unspecified remainder-solution of the compartment will be split equally during rule instantiation and the resulting sub-solutions will be bound to the according variables. These variables are reused on the right-hand side of the rule to distribute the split content among both daughter cells. Please notice, the solution that is enclosed by species **Mem** is decomposed into three parts m_b?, m_i? and m_j?. Thereby, decomposition is constrained, such that sub-solution m_b? consists of all species ?($state^{x \neq F}$), i.e., of all non-free Notch and Delta molecules. The remaining part is split equally again and the resulting sub-solutions are bound to variables m_i? and m_j?. This constrained decomposition prevents from inserting receptors and ligand molecules into the newly created cell, which are in fact tightly connected to the dividing cell due to intercellular bonds.

The entire Notch signaling model encoded in the basic textual syntax of ML-Rules can be found in Figure 8.2, which describes the dynamics within and between arbitrarily many cells and thereby demonstrates the succinctness of the developed modeling approach. A typical simulation study with this model could investigate under which conditions cell proliferation will stop due to lateral inhibition. For instance, we would expect that cells with only one other cell in their adjacency consist of fewer activated Notch receptors than cells with two neighbors. Larger amounts of activated Notch, in turn, might lead to an increased translocation of Nicd into the nucleus, resulting in a higher expression level of newly synthesized Notch and subsequently also in more receptor molecules attached to the cell membrane. Receptor activation may thus initiate a positive feedback loop, such that a cell becomes more and more activated by increasing amounts of Notch. In addition, larger amounts of cytoplasmic Notch are assumed to cause an inhibition of cell proliferation, i.e., the cell will not divide as it becomes arrested in the quiescent cell cycle phase G_0. As this process of lateral inhibition plays an important role in many systems, e.g., in epidermal regeneration during wound healing, an investigation of different subcellular, cellular, or cell population conditions regarding their effects on expected or abnormal behavior might reveal new insights.

However, the purpose of this toy example is to demonstrate – within a single model – diverse aspects that are relevant for multilevel modeling in systems biology. Therefore, some artificial assumptions have been introduced, e.g., it is not clear whether the growing membrane volume really plays an important role for the rate of receptor activation. Other important regulatory dynamics have been left away to keep the model simple, e.g., the processes of transcriptional activation and inhibition are highly simplified. For this reasons, simulation results of the Notch signaling model are not presented here. However, in addition to model encodings, the second case study on fission yeast cell proliferation involves also some simulation experiments.

```
// species definitions
Cell(2);
Mem(1);
Cyt(1);
Nuc(1);
N(1);
D(1);
FS(1);

// initial solution
>>INIT[ for i:min while (i<0)  with i+1 [ FS(i) ] +
        Cell(0,'G1')[Mem(Vm0)[Cyt(Vc0)[Nuc(0)]]]  +
        for i:0 while (i<=max) with i+1 [ FS(i) ]
    ];

// synthesis of Notch and Delta
Nuc(n)[s?] -> Nuc(n)[s?] + N('F') @ k1prime+(k1*((n^h)/(K^h + n^h)));
Nuc(n)[s?] -> Nuc(n)[s?] + D('F') @ k2*(1-((n^h)/(K^h + n^h)));

// protein translocation to membrane
Cyt(v)[N('F'):n + s?] -> Cyt(v)[s?] + N('F') @ k3*#n;
Cyt(v)[D('F'):d + s?] -> Cyt(v)[s?] + D('F') @ k4*#d;

// receptor-ligand binding
Cell(i,pi)[Mem(vi)[N('F'):n + mi?]] + Cell(j,pj)[Mem(vj)[D('F'):d + mj?]] -> Cell(i,pi)[Mem(vi)[N(j) +
        mi?]] + Cell(j,pj)[Mem(vj)[D(i) + mj?]] @ if (j==i+1) || (j==i-1) then (k5/(vi+vj))*#n*#d else 0;

// catalysed cleavage of activated Notch
Cell(i,pi)[Mem(vmi)[Cyt(vci)[ci?] + N(j):n + mi?]] + Cell(j,pj)[Mem(vmj)[D(i) + mj?]] ->
        Cell(i,pi)[Mem(vmi)[Cyt(vci)[N('I') + ci?] + mi?]] + Cell(j,pj)[Mem(vmj)[D('E') + mj?]] @
        (vmax*#n)/(Km+#n);

// translocation of Nicd into the nucleus
Nuc(n)[s?] + N('I'):c -> Nuc(n+1)[s?] @ k7*#c;

// recycling of Delta ligand
Cyt(v)[s?] + D('E'):d -> Cyt(v)[D('F') + s?] @ k8*#d;

// protein degradation
N(state):n -> @ if (state=='F') || (state=='I') then kdeg*#n else 0;
D(state):d -> @ if (state=='F') || (state=='E') then kdeg*#d else 0;
Nuc(n)[s?] -> Nuc(n-1)[s?] @ if (n>0) then kdeg*n else 0;

// cell growth
Mem(vm)[Cyt(vc)[c?] + m?] -> Mem(vm+(dg/2))[Cyt(vc+dg)[c?] + m?] @ if (vc<(2*Vc0)) then kg*vc else 0;

// cell cycle phase transition (G1 checkpoint)
Cell(i,'G1')[Mem(vm)[Cyt(vc)[N('F'):n + c?] + m?]] -> Cell(i,if (#n<tN) then 'G2' else
        'G0')[Mem(vm)[Cyt(vc)[N('F') + c?] + m?]] @ if (vc>=(1.5*Vc0)) then kG1 else 0;

// cell division
Cell(i,'G2')[Mem(vm)[Cyt(vc)[c?] + m?]] + FS(j) -> Cell(i,'G1')[Mem(Vm0)[Cyt(Vc0)[c?] + m?]] +
        Cell(j,'G1')[Mem(Vm0)[Cyt(Vc0)[Nuc(0)]]] @ if (vc>=2*Vc0) && ((j==i+1)||(j==i-1)) then kG2 else 0;
```

Figure 8.2: ML-Rules model of the Notch signaling example. Parameter definitions are omitted.

8.2 Fission Yeast Cell Proliferation Model

8.2.1 Main Properties and Structure of the Model

The second case study is a multilevel model of the fission yeast (*Schizo-saccharomyces pombe*) cell proliferation in dependence on an intracellular control circuit. The dynamics of this intracellular regulatory network of interacting proteins in turn depends on the cell's size and the concentration of specific pheromone molecules within its environment that play an important role in sexual reproduction. To prepare for mating, fission yeast secretes these pheromones, which subsequently may cause an arrest of the division cycle of cells with an opposite mating type. Different cells may thereby communicate with each other over rather long distances due to pheromone diffusion. Hence, the model's dynamics operate at multiple levels of organization and the different parts are highly interconnected and influence each other in various ways.

In the following sections, a rather simple single-cell model of the fission yeast cell cycle regulation will be successively extended to a hierarchically organized multicellular model. An overview of this model's structure and dynamics is depicted in Figure 8.3. The model comprises three distinct hierarchical levels. At the bottom level, interacting proteins describe the intracellular dynamics of a fission yeast cell. The intermediate level describes dynamics at the level of entire cells, i.e., cell growth, cell cycle phase transitions, and cell division including mating type switching. Also at this hierarchy level, pheromone molecules are secreted into the extracellular medium. Various interlevel causalities between the intermediate and the bottom level influence dynamic processes both in an upward causation (processes 7–9) as well as downward causation manner (4,11–12). In addition, the top level discretizes the environment of cells into multiple fictive compartments to cover diffusion of pheromones and excluded volume effects leading to the displacement of cells from crowded to more empty areas.

8.2.2 Downward Causation in Cell Cycle Regulation

As has been already outlined in Chapter 3, the eukaryotic cell cycle consists of four phases: G_1, S, G_2, and M. During the first three phases, the cell is

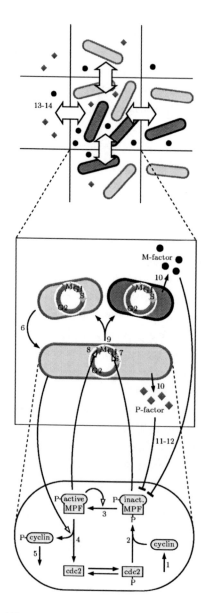

Figure 8.3: Overview of the multilevel yeast cell proliferation model. (Reactions 1–5) intracellular dynamics, (6) cell growth, (7–8) cell cycle phase transitions, (9) cell division and mating type switching, (10) secretion of pheromone (P-factor and M-factor) molecules, (11–12) pheromone signaling, (13) pheromone diffusion, (14) spatial displacement of cells.

increasing in size and its DNA is replicated. At the end of the cycle, the cell enters the M phase (mitosis) and finally divides into two daughter (or sibling) cells. These major events of the cell division cycle are controlled by certain proteins (primarily cyclins and cyclin-dependent kinases) and the underlying processes in fission yeast have been extensively studied in the past decades (Kim and Huberman, 2001; Tyson and Novak, 2001; Tyson et al., 2002; Kar et al., 2009; Moseley et al., 2009; Sawin, 2009).

The low-level regulating processes in our case study are based on an early model by John Tyson (1991). This model consists of two proteins (cyclin and cdc2) that form a complex called maturation promoting factor (MPF), which in turn may be activated by an autocatalytic process and disassembles afterwards. Thereby, the varying amounts of inactive and activated MPF are assumed to control passing through the cell cycle, e.g., entering the M phase requires large amounts of active MPF. The according biochemical reaction network is illustrated by green colored entities at the bottom of Figure 8.3. Although today much more detailed models of cell cycle regulation exist (e.g., Novak and Tyson, 1997; Sveiczer et al., 2001; Qu et al., 2003; Kar et al., 2009), the Tyson model denotes a suitable starting point for studying cell cycle regulation, as it is simple but at the same time captures the essential dynamics. Moreover, the model comprises some implicit notion of downward causation and is therefore well qualified for our use case.

Tyson identified specific regions in the parameter space – in particular for the rate coefficients of autocatalytic MPF activation and its disassembly – where regular cycle oscillations with bursts of the amounts of inactive and activated MPF can be observed. However, with constant rate coefficients, the period of roughly 30 minutes between two peaks is significantly shorter than the mean mass-doubling time of wild type fission yeast of 116 minutes (Miyata et al., 1978). Therefore, to achieve longer oscillation periods, Tyson assumes with increasing cell size a dilution of an enzyme catalyzing the breakage of MPF into cdc2 and cyclin-P and the according rate coefficient is thus decreased during the cycle. Hence, implicitly, the Tyson model already includes dynamics at different organizational levels as well as downward causation from the cellular level to a process at the subcellular level. These multiple levels

Table 8.2: Species definitions of a simple cell cycle regulation model.

Biological structure / entity	Species name	Attribute names
Fission yeast cell	**Cell**	(*vol*)
Cdc2	**Cdc2**	( )
Cyclin	**Cyclin**	( )
Phosphorylated cyclin	**CyclinP**	( )
Inactive MPF complex	**MPFi**	( )
Active MPF complex	**MPFa**	( )

and their interrelation shall be made explicit now. Therefore, in addition to different protein species, also an attributed species **Cell**(*vol*) is specified to describe the cell compartment and its size in terms of volume (see Table 8.2). By doing so, the rate of MPF disassembly inside the cell can be dynamically adjusted by taking the current volume of the cell into account.

Please notice, unlike it is done for cyclin, we do not distinguish between phosphorylated and unphosphorylated cdc2. This is a simplification in accordance with the original model, as the phosphorylation and dephosphorylation reactions of cdc2 are very fast compared to the others and can thus be neglected. Also, although the different states of cyclin and MPF may be straightforwardly represented by according attribute values rather than distinct species names, here the species have not been equipped with attributes since each of them may have two different states only and in this case the specification of rules becomes more compact than it would be possible by using attributes.

The initial solution of the model consists of only one **Cell** species, as here the interplay between cell growth and intracellular biochemical reactions shall be investigated. An extension of the model describing the behavior of multiple cells will be discussed later. Enclosed by the **Cell** – whose volume attribute *vol* is assigned the value **1** – is a number of $cdc2_{tot}$ species **Cdc2**. Other species are not part of the initial solution:

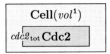

The first reaction rule describes the zero-order synthesis of cyclin, added to the solution enclosed be the cell:

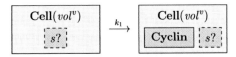

Please note, since the model does not comprise more than one **Cell** species so far, there is no need to consider its amount in the firing rate of the rule. Hence, there is also no need for a species identifier of the **Cell** reactant, like it is done in the following rule, which describes the formation of the inactive MPF complex by an association of cyclin and cdc2.

$$\boxed{\text{Cyclin}}^y\boxed{\text{Cdc2}}^d \xrightarrow{k_2\#y\#d} \boxed{\text{MPFi}}$$

As has been stated earlier, activation of inactive MPF – i.e., dephosphorylation of its cdc2 subunit – is assumed to be an autocatalytic process. That means, the higher the amount of activated MPF, the higher is the activation rate. Therefore, we need to access the amount of **MPFa**, which is done by using a counting function on the bound solution $s?$:

$$\boxed{\begin{matrix}\text{Cell}(vol^v)\\[2pt]\boxed{\text{MPFi}}^m\,\lfloor s?\rfloor\end{matrix}} \xrightarrow{k_3'+k_3\left(\frac{\sum\text{MPFa}\in s?}{cdc2_{tot}}\right)^2\#m} \boxed{\begin{matrix}\text{Cell}(vol^v)\\[2pt]\boxed{\text{MPFa}}\,\lfloor s?\rfloor\end{matrix}}$$

Alternatively, to avoid an application of the counting function, one could introduce an additional cell attribute denoting the current amount of enclosed **MPFa**, similar to what has been done for accessing the amount of nuclear Nicd in the Notch signaling model (cf. page 182f). However, due to the specific rate law of the MPF activation process, here it would be sufficient to split it up into a *basal* and a *catalyzed* activation process. The according rules are as follows:

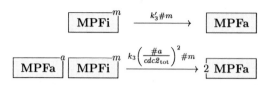

The rule schema below describes the MPF disassembly process, whose rate is assumed to decrease with growing cell size. Therefore, the rate of the rule is dynamically adjusted in a downward causation manner by taking the cell's current compartment volume into account:

$$\boxed{\begin{array}{c} \text{Cell}(vol^v) \\ \boxed{\text{MPFa}}^{\,m} \boxed{s?} \end{array}} \xrightarrow{(k_4/v)\#m} \boxed{\begin{array}{c} \text{Cell}(vol^v) \\ \boxed{\text{CyclinP}}\ \boxed{\text{Cdc2}}\ \boxed{s?} \end{array}}$$

Subsequent degradation of phosphorylated cyclin is described by an ordinary first-order reaction with mass-action kinetics:

$$\boxed{\text{CyclinP}}^{\,y} \xrightarrow{k_5\#y} \emptyset$$

The rules so far describe the intracellular dynamics of the cell cycle regulation model. What is still missing are the rules for cell growth, i.e., an increase of the volume over time, and the abrupt reduction of the volume that mimics cell division. Therefore, the growth process is discretized such that the volume attribute of the **Cell** species is increased by $1/T_d$ with a rate constant $k_6 = 1$ min^{-1} and the mean doubling time T_d given in minutes:

$$\boxed{\begin{array}{c} \text{Cell}(vol^v) \\ \boxed{s?} \end{array}} \xrightarrow[v<2]{k_6} \boxed{\begin{array}{c} \text{Cell}(vol^{v+1/T_d}) \\ \boxed{s?} \end{array}}$$

The cell volume in this model is represented by relative rather than absolute values, where 1 and 2 are the volumes at birth and division respectively. That is why the above rule for cell growth is constrained to only fire as long as the volume is lower than 2. Consequently, the cell division process – where the cell's volume gets halved – happens after the volume is greater or equal 2:

$$\boxed{\begin{array}{c} \text{Cell}(vol^v) \\ \boxed{s?} \end{array}} \xrightarrow[v\geq 2]{k_9} \boxed{\begin{array}{c} \text{Cell}(vol^{v/2}) \\ \boxed{s?} \end{array}}$$

Please notice that a firing of this rule does not increase the number of cells in the model, although this could be easily expressed by putting two cells on the right-hand side of the rule. Later, the model will be extended this way. However, at this stage the model shall describe only the single-cell interplay between subcellular processes and dynamics at the cellular level. Therefore, the cell number and its enclosed content are kept constant and the above rule merely mimics cell division by reducing the cell's volume.

Unlike the continuous deterministic analysis by Tyson (1991), the stochastic simulation of the presented ML-Rules model (see Figure 8.4) reveals highly variable oscillation periods as well as nonconformity between the time points of active MPA bursts and cell division events, i.e., an abrupt decrease of the cell's volume. Hence, to get more lifelike oscillatory behavior in the presence of stochastic effects and to achieve better accordance between protein peaks and division times, the presented model needs some modifications. Let us in the next section therefore take a look on how low-level states influence dynamics at the cell level, i.e., how intracellular dynamics trigger high-level events so that the cell traverses through the different phases of the cell cycle.

8.2.3 Controlling Cell Proliferation via Bidirectional Interlevel Causation

The accumulation of inactive MPF, i.e., the protein complex where both subunits cyclin and cdc2 are phosphorylated, denotes the initiation of DNA synthesis. Hence, inactive MPF controls the transition from G_1 to S phase of the cell cycle. Similarly, the transition from G_2 to M phase and the final division process are controlled by the rapid increase and subsequent depletion of the amount of active MPF. Therefore, cell cycle phase transitions and thus also the duration of the cycle are influenced by dynamics at a lower level of organization.

To describe these upward causalities, the **Cell** species is equipped with an additional attribute (*phase*) that denotes the current phase of the cell cycle. In the initial solution, it is assigned the value G1. As DNA replication – i.e., the S phase – takes a rather constant amount of time and here the interest lies in the control of G_1 and G_2 checkpoints, the S and G_2 phases of the cell cycle

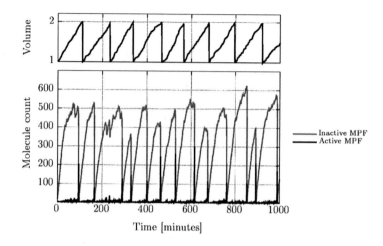

```
// parameters
cdc2tot:   1000;
Td:        116;
k1:        0.015*cdc2tot;
k2:        200;
k3:        180;
k3prime:   0.018;
k4:        4.5;
k5:        0.6;
k6:        1.0;
k9:        1e6;
```

```
// species definitions
Cell(1);
Cdc2();
Cyclin();
CyclinP();
MPFi();
MPFa();

// initial solution
>>INIT[ Cell(1)[cdc2tot Cdc2] ];
```

```
// rule schemata
Cell(v)[s?]                  -> Cell(v)[Cyclin + s?]                  @ k1;
Cyclin:y + Cdc2:d            -> MPFi                                  @ k2*#y*#d;
MPFi:m                       -> MPFa                                  @ k3prime*#m;
MPFa:a + MPFi:m              -> 2 MPFa                                @ k3*((#a/cdc2tot)^2)*#m;
Cell(v)[MPFa:m + s?]         -> Cell(v)[CyclinP + Cdc2 + s?]          @ (k4/v)*#m;
CyclinP:y                    ->                                       @ k5*#y;
Cell(v)[s?]                  -> Cell(v+(1/Td))[s?]                    @ if (v<2) then k6 else 0;
Cell(v)[s?]                  -> Cell(v/2)[s?]                         @ if (v>=2) then k9 else 0;
```

Figure 8.4: ML-Rules variant of the Tyson (1991) model of yeast cell cycle regulation. The model includes downward causation by dynamic adjustment of the MPF disassembly rate due to an increase of the cell's volume. (Top) Stochastic simulation results. (Bottom) Model encoding in the basic textual ML-Rules syntax.

are combined to a single phase S/G_2. The transition from phase G_1 to phase S/G_2 is described as follows:

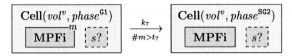

where t_7 denotes a certain threshold value of inactive MPF that needs to be exceeded in order to pass the G_1 checkpoint. Similarly, the transition from S/G_2 to M phase is guarded by a threshold t_8 in the amount of active MPF:

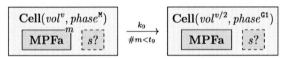

The last cell cycle phase transition from M back to G_1 denotes the division into two daughter cells. However, we are still interested in the interplay between high-level and low-level states only. Thus, the number of cells is kept constant and only the volume of the "dividing" cell gets reduced, like is has been done in the previous section. Cell division occurs after the amount of active MPF falls below a second threshold value t_9, which is significantly lower than t_8:

Please note, in all three phase transition rules the amount of an enclosed species needs to be accessed in order to determine whether the rule may fire or not. This is achieved here by explicitly specifying the respective species on the left-hand side of the rule (to prevent from consumption also on the right-hand side), which allows for assigning a species identifier and thereby for retrieving its current amount. However, alternatively one could also use a counting function on the bound solution $s?$, e.g., $\sum \mathbf{MPFa} \in s?$.

Now that the model comprises defined states for the different phases of the division cycle, the cell must no longer be restricted to grow in size until its volume has doubled. Instead, the volume is allowed to increase at any time but not during the M phase:

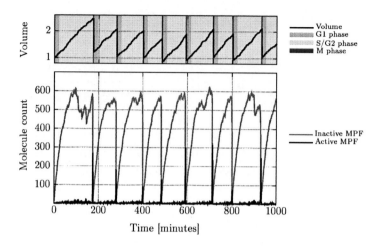

Figure 8.5: Yeast cell cycle regulation model comprising downward and upward causation. MPF disassembly depends on the cell's volume and at the same time, the cell cycle duration depends on the intracellular amounts of active and inactive MPF. The concrete model encoding (including parameters) can be found in Appendix B.1.

$$\text{Cell}(vol^v, phase^p) \quad \xrightarrow[p \in \{\texttt{G1,SG2}\}]{k_6} \quad \text{Cell}(vol^{v+1/T_d}, phase^p)$$

Compared to the original model with downward causation only, simulating the extended multilevel model – that comprises also upward causation – shows rather stable oscillation periods in accordance with the mean mass-doubling time T_d of 116 minutes (Figure 8.5). Cell division may happen at volumes larger than 2 and consequently the cell cycle may take more time than T_d, but if this is the case, this has implications for the next cycle: Due to the unusually large volume at birth, MPA activation then occurs relatively fast and thus the following cycle tends to be shorter than normal. In this way, the combination of upward causation and downward causation regulates both, the cell cycle duration and cell size homeostasis.

The results emphasize the role of multiple levels and their inter-relation in studying phenomena like cell division. They also show the necessity for flexibly constraining dynamic processes to describe interlevel causalities. Therefore, ML-Rules provides nesting of species and rates with arbitrary kinetics based on constraints. So far, the model appears still to be manageable in other – less expressive – modeling languages. However, the next step, i.e., moving from a single-cell model to multicellular dynamics, illuminates the importance to be able to express multiple levels and interlevel causation explicitly, succinctly, and in a flexible manner.

8.2.4 Cell Division and Mating Type Switching

Besides asexual cell division, the unicellular fission yeast may also undergo sexual reproduction when environmental conditions are getting poor, e.g., when cells are starving. Different mating types (P and M) exist enforcing fusion of cells of opposite types only (Leupold, 1958). The product of fusion is a diploid zygote rapidly entering a sporulation process. Later, when the environmental conditions improve, spores germinate to spawn haploid cells, which then undergo normal asexual proliferation again. The mating type of proliferating cells switches sporadically when a cell divides. This phenomenon is regulated by rather complex gene-regulatory mechanisms (Beach, 1983; Beach and Klar, 1984; Egel, 1984a,b; Klar et al., 1991; Yamada-Inagawa et al., 2007). However, rather robust phenomenological switching patterns can be observed (Klar, 1992). One important characteristic is that cells do not only show one of the two different mating types, i.e., P or M, but can be also categorized into cells that are able to switch their type and those that are not (Figure 8.6).

Although comprising multiple levels, the cell proliferation model so far describes the dynamics of a single cell only. In order to describe a multicellular system, we extend the previous cell division rule – i.e., the cell cycle transition from phase M to phase G_1 – such that the dividing **Cell** species is replaced by two **Cell** species on the product-side of the rule. At the same time, instead of modeling detailed processes at the genetic level, simple phenomenological alterations of the cell's mating type are assigned according to the regularities depicted in Figure 8.6. Therefore, the **Cell** species is equipped with two additional attributes: *type* for representing the cell's mating type and *sw* to denote

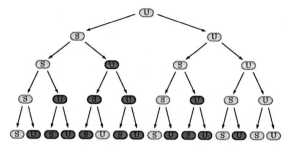

Figure 8.6: Switching of mating types in a fission yeast cell lineage. Cells of type M are marked by dark shadings, light gray stands for mating type P. The unswitchable and switchable states are denoted by U and S respectively. Reproduced figure from Klar (1992).

whether the cell is able to switch its type or not. This leads to the following rule schema describing the division of an unswitchable cell:

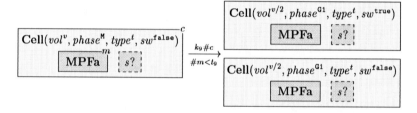

The complementary schema for the division of switchable cells looks pretty much similar:

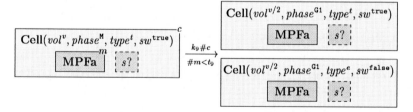

with conditional expression $e :=$ if $t = $ P then M else P.

Please notice, since the former single-cell model has now been extended to describe multicellular dynamics and therefore populations may exist that comprise multiple identical **Cell** species, here also the cell amount needs to be taken into account in the kinetic rate expression. The multicellular model, in which each cell may have its own state and behavior, also underlines the necessity for specifying rule *schemata*. Otherwise, modeling such systems would be highly impractical if not impossible, as each defined combination of cellular attributes and enclosed content would require the specification of an own rule.

The above rule schemata also illustrate the importance for binding unspecified solutions to variables, so that arbitrary content of species can not only be preserved but also copied and flexibly inserted into multiple species on the right-hand side of a rule. Like in the previous single-cell model, the solution here also remains unsplit and is thus not distributed among both daughter cells. The contained solution will be entirely copied instead, as in accordance to the original Tyson model the total amount of cdc2 protein, i.e., the sum of the amounts of species **Cdc2**, **MPFi**, and **MPFa**, is assumed to be constant in each cell. However, by applying the universal decomposition approach introduced in Chapter 7.4.3, it would be also possible to split the solution according to certain constraints. For instance, the rule below describes an equal splitting of the solution. In addition, this rule also shows an application of the counting function that aggregates the amounts of **MPFa** within different solutions:

$$\textbf{Cell}(vol^v, phase^{\texttt{M}}, type^t, sw^{\texttt{false}})[s_1? + s_2?]^c$$

$$\xrightarrow[(\sum \textbf{MPFa} \in s_1? + s_2?) < t_9]{k_9 \# c}$$

$$\textbf{Cell}(vol^{v/2}, phase^{\texttt{G1}}, type^t, sw^{\texttt{true}})[s_1?] \; + \; \textbf{Cell}(vol^{v/2}, phase^{\texttt{G1}}, type^t, sw^{\texttt{false}})[s_2?]$$

As shown by the simulation results in Figure 8.7, mating type switching ensures that – in the long run – both types P and M are equally distributed in a population of fission yeast cells. The initial population consisting of only one type of cells reveals distinct time points of the first appearance of cells comprising other combinations of mating type and the ability to switch. Also an equilibration of the different mating types after just a few division cycles can be observed. However, at this stage, different cells in the model are not interacting with each other, as the dynamics of a cell only depends on its own state and on the state of its enclosed solution. In the next step, this model will be extended once again to also describe intercellular communication.

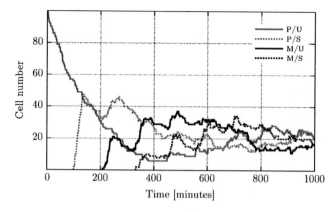

Figure 8.7: Multicellular model of yeast proliferation and mating type switching. Trajectories of different cell types from a stochastic simulation run with an initial population of 100 unswitchable cells of mating type P. The concrete model encoding (including all parameters) can be found in Appendix B.2.

8.2.5 Intercellular Communication Through Pheromones

Conjugation of two fission yeast cells is restricted to poor environmental conditions and to cells of opposite mating types. To only prepare for mating in the presence of compatible partners, fission yeast cells communicate with each other through diffusible pheromone molecules (Nielsen and Davey, 1995). The fission yeast pheromone secretion and response is therefore an example of paracrine signaling. When growing in a nitrogen-poor environment, for instance, cells are starving and begin to synthesize mating type specific pheromones that are getting secreted to the extracellular medium. Thereby, the pheromone that is secreted by cells of mating type P is called P-factor and M-type cells produce the M-factor pheromone. Sensing of pheromone molecules released by the opposite type causes several regulation processes that prepare the cells for mating. One of the main effects is an arrest of the cell division cycle in the G_1 phase (Stern and Nurse, 1997). In this section, the yeast proliferation model will be extended by adding certain species and rules to describe communication via pheromone molecules and the respective response leading to a G_1 arrest. Thereby, intercellular communication relies not on direct cell-to-cell interaction (juxtacrine signaling) like in the previous example of Notch signaling, but on diffusible molecules.

Therefore, at first additional rules for pheromone secretion and degradation are specified, where **Phe**($type$) denotes an additionally defined attributed species name that – depending on the attribute's value – represents P-factor and M-factor pheromone respectively:

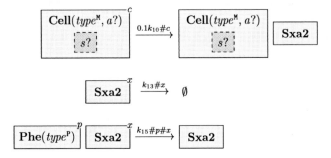

Please note, the pheromone secretion rule takes advantage of identifying attributes by name, as this method allows to leave certain attributes unspecified if they are of no interest – here the *vol*, *phase*, and *sw* attributes of species **Cell** – and thereby notations become more succinct.

Once secreted, M-factor molecules, i.e., species **Phe**($type^M$), may influence dynamics of P-type cells. Conversely, cells of mating type P may communicate with M-type cells via P-factor molecules, i.e., via **Phe**($type^P$). In addition to M-factor, cells of type M also produce and release a P-factor-specific protease (Sxa2), which catalyzes the degradation of P-factor and thereby lowers the effect that P-type cells have on M cells (Imai and Yamamoto, 1992; Nielsen and Davey, 1995):

For the sake of simplicity, instead of modeling a detailed pheromone response signaling cascade including receptor binding, here the amount of extracellular pheromone is simply taken into account for influencing the dynamics

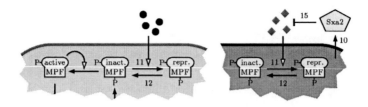

Figure 8.8: Abstract model of MPF repression in dependence on extracellular pheromone. M-factor molecules (black circles) have an effect on cells of mating P, while P-factor (gray diamonds) only influences the dynamics of M-type cells. P-factor pheromone is catalytically degraded by Sxa2, which is secreted by cells of mating type M only.

of the intracellular control circuit. As already mentioned, the presence of pheromone may cause a G_1 arrest of the cell cycle. It has been shown that inhibition of the cyclin-cdc2 complex is crucial for this process (Stern and Nurse, 1997). Therefore, a new species name **MPFr** denoting a repressed MPF complex is introduced that prevents inactive MPF from being activated. However, a repressed MPF complex may spontaneously recover to its normal inactive state (see also Figure 8.8). The reaction rate of MPF repression is assumed to depend on the amount of environmental pheromone. Since extracellular pheromone resides at a higher nesting level than the intracellular MPF complex, the repression process is constrained by downward causation. However, different from previous examples, here the interlevel causation acts across the boundaries of a nested species, i.e., the **Cell**, and not just between an attributed species and its enclosed solution:

$$\boxed{\mathbf{Phe}(type^t)}^p \begin{bmatrix} \boxed{\mathbf{Cell}(type^e, a?)}^c \\ \boxed{\mathbf{MPFi}}^m \ \boxed{s?} \end{bmatrix} \quad \left(\frac{k_{11}\#p^3}{K_{11}^3+\#p^3} \frac{1}{v^2} \right) \#m\#c \quad \boxed{\mathbf{Phe}(type^t)} \begin{bmatrix} \boxed{\mathbf{Cell}(type^e, a?)} \\ \boxed{\mathbf{MPFr}} \ \boxed{s?} \end{bmatrix}$$

with conditional expression $e :=$ if $t = $ P then M else P.

In fact the above rule of pheromone-induced MPF inhibition includes two different downward causalities at the same time. The first is the amount of extracellular pheromone, which is included in the rate factor that describes a Hill type sigmoidal response curve for MPF repression. The second downward causation is a volume-dependence again. This reflects the observation that inhibition of MPF activity is partly lost due to increasing cell size Stern and Nurse (1997), which could be, for instance, again the consequence from a dilution of involved but here not explicitly considered enzymes. Finally, recovery of inactive MPF from its repressed state is described as follows:

$$\boxed{\text{MPFr}}^{\overline{-m}} \xrightarrow{k_{12}\#m} \boxed{\text{MPFi}}$$

The simulation results given in Figure 8.9 show how the additional reaction rules influence the intracellular processes and by that have an effect on the dynamics at cell level, i.e., progression through the cell cycle phases. As pheromone secretion and mating takes place when nutrition is poor, the parameter T_d denoting the cell's mass-doubling time has been increased to 232 minutes in these simulation experiments. In the absence of pheromone, the cell cycle length then increases to roughly 200 minutes (Figure 8.9a). Similar dynamics can be observed in the presence of a low amount of extracellular pheromone, as the pheromone's effect on MPF activation through repression is too low (Figure 8.9b). This is different with a higher pheromone concentration: Figure 8.9c indicates a strong suppression of inactive MPF by the repressed variant. However, repression gets partly lost over time, i.e., the cell adapts while it grows in size and completes the cell cycle finally after more than 600 minutes. Please notice the dramatically increased duration of the G_1 phase, while both the S/G_2 and M phase take only slightly more time than without pheromone signaling.

Compared with an exponential population growth unaffected by any pheromones, the simulation of a multicellular model comprising intercellular communication shows significantly reduced cell numbers due to an arrest in the G_1 phase and thus an increased mean cell cycle duration (Figure 8.10). With an increasing cell number, the pheromone amounts increase as well and thereby larger fractions of cells being in the G_1 phase can be observed at later time points. After 1000 minutes simulation time, nearly half the population is arrested in the G_1 phase. At this point, the amounts of pheromones lie be-

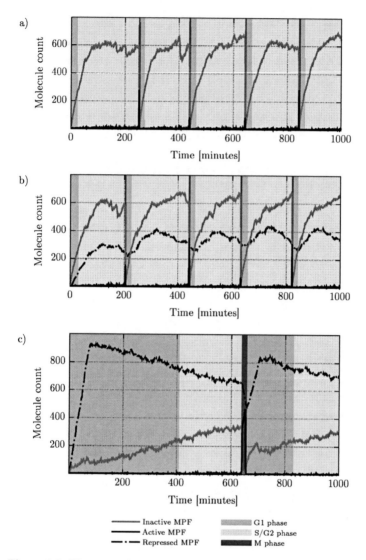

Figure 8.9: Pheromone-dependent cell cycle dynamics of a single fission yeast cell. MPF trajectories and cell cycle phase durations with pheromone different amounts: (a) in the absence of pheromone, (b) with an extracellular pheromone amount of 200 molecules, and (c) in the presence of 600 molecules. The model encoding can be found in Appendix B.3.

tween 400 (P-factor) and 800 (M-factor) molecules. However, although the P-factor-specific protease Sxa2 lowers the amount of P-factor pheromone and thus lowers the signaling effects on M cells, mating type switching still ensures equal distributions of the different mating types P and M.

The presented extension of the previous proliferation model of non-communicating cells to a model comprising intercellular communication through paracrine signaling illustrates the usefulness of compositionality, as this allowed us to simply add a few new species and rules while leaving the remaining model parts completely unmodified. The model also underlines again the importance of constraining reactions flexibly.

The model so far assumes all cells and secreted pheromones residing in the same well-mixed solution, i.e., there is no spatial information included. However, since pheromone molecules may diffuse within the medium after secretion and also individual cells of a population may be inhomogeneously distributed in space, the investigation of spatial dynamics might be desired. Therefore, in the next section the model will be extended once more time to also cover pheromone diffusion and different locations of cells.

8.2.6 Spatial Layer

To investigate the relation between signaling, pheromone diffusion, and the location of cells, we adopt the idea of Elf and Ehrenberg (2004) and discretize the model's space into multiple subvolumes. Therefore, an additional attributed species $\mathbf{SV}(x, y)$ is defined to represent virtual reaction compartments. Attributes x and y thereby represent the spatial coordinates of a subvolume within a two-dimensional lattice. Hence, the initial solution comprises $x_{max} \times y_{max}$ species $\mathbf{SV}$, each with a unique combination of attribute values x and y with $x \in \{1, \ldots, x_{max}\}$ and $y \in \{1, \ldots, y_{max}\}$ (see also Figure 8.11). Each subvolume may then enclose a solution consisting of cells and pheromone molecules. Like before, all species within such a solution are assumed to be homogeneously distributed, however, since species may migrate to adjacent subvolumes according certain rules, different spatial locations of cells and pheromones can be distinguished from each other in a coarse-grained manner.

Diffusion of molecules can be described by a simple movement from one

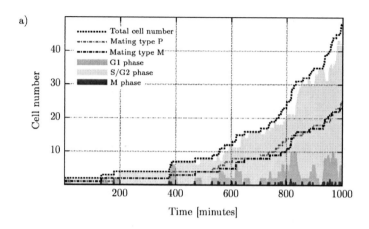

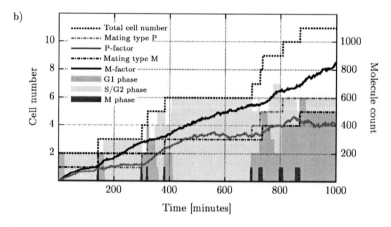

Figure 8.10: Cell population growth and intercellular communication. Simulation results of the multicellular model that combines cell division with mating type switching and pheromone response. (a) Exponential population growth in the absence of pheromone secretion and response. (b) Pheromone secretion and response leads to G_1 arrest and reduced population growth rate. The model encoding including parameters for both simulation experiments can be found in Appendix B.4.

$$\text{SV}(x^1, y^2) \quad \text{SV}(x^2, y^2) \quad \text{SV}(x^3, y^2) \quad \text{SV}(x^4, y^2)$$

$$\text{SV}(x^1, y^1) \quad \text{SV}(x^2, y^1) \quad \text{SV}(x^3, y^1) \quad \text{SV}(x^4, y^1)$$

Figure 8.11: Initial solution of a 4×2 lattice of discrete subvolumes.

subvolume to an adjacent one. By comparing the coordinates of two species **SV**, the diffusion can be constrained to take place only between neighboring locations. Hence, the rule schema for pheromone diffusion in a von Neumann neighborhood is described as follows:

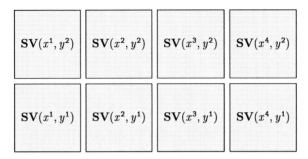

with expression $e_{ij} := (x_i = x_j \wedge (y_i = y_j + 1 \vee y_i = y_j - 1)) \vee (y_i = y_j \wedge (x_i = x_j + 1 \vee x_i = x_j - 1))$ constraining the diffusion to adjacent subvolumes.

The former pheromone degradation rule is replaced here by rules describing diffusion out of the system, i.e., for diffusing particles the lattice is assumed to be an open system. Therefore, the following rules check whether a certain subvolume is located at the boundary of the lattice. If this is the case, a pheromone molecule is simply removed with a certain probability.

$$\text{SV}(x^x, y^y) \quad \text{Phe}(type^t)^{p} \quad s? \xrightarrow[\ x=1 \vee x=x_{max}\ ]{\frac{1}{4}k_{13}\#p} \quad \text{SV}(x^x, y^y) \quad s?$$

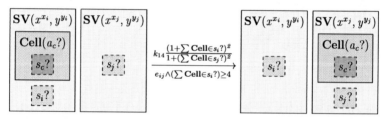

Unlike the diffusing pheromone molecules, in this model cells are not assumed to move randomly in space. Instead, we describe some kind of *excluded volume effect* to avoid that too many cells occupy one subvolume. Therefore, if a location gets crowded, cells may be "pushed" to an adjacent and less crowded subvolume. The rule for such a displacement from crowded areas looks quite similar to the rule that describes diffusion. Certain constraints restrict the movement to neighboring subvolumes and the kinetic rate depends on the number of cells within the subvolumes to specify the probability with which a cell may move to an adjacent location. The rate of the following rule, for instance, makes migration to empty locations more likely than those to crowded ones:

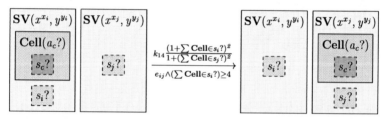

with expression $e_{ij} := (x_i = x_j \wedge (y_i = y_j + 1 \vee y_i = y_j - 1)) \vee (y_i = y_j \wedge (x_i = x_j + 1 \vee x_i = x_j - 1))$ constraining the migration to an adjacent subvolume.

Please notice, since the cells in our model typically denote distinct species with different combinations of attribute values and enclosed content, it is not possible to use species identifiers to get the total number of cells within a solution. Instead, the above rule employs a counting function on the bound solutions $s_i?$ and $s_j?$. Alternatively, the **SV** species may be equipped with an additional attribute that holds the current total number of cells in each subvolume. However, in this case, any rule that creates or removes a **Cell** species, e.g., a cell division rule, then needs to be extended such that its context includes the high-level **SV** species for updating the according attribute. This

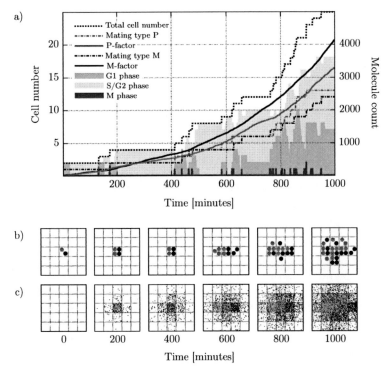

Figure 8.12: Simulation results of the spatial yeast cell proliferation model. The model encoding (including parameters) can be found in Appendix B.5. (a) Trajectories of the overall cell population and pheromone molecules. (b) Scatter plots of the spatial distribution of cells at six distinct time points. The different mating types P and M are depicted by gray and black circles respectively. (c) Scatter plots of spatial pheromone distribution. Gray and black dots denote P-factor and M-factor molecules respectively.

denotes an undesired artifact and emphasizes the importance of supporting functions on solutions again.

Simulating yeast cell proliferation within the described spatial setting reveals that although local differences in the amounts of P and M type cells can be observed, the overall ratio between the different mating types remains nearly constant over time (Figure 8.12). Secreted pheromones cause large fractions of cells to be arrested in the G_1 phase. However, the overall cell population is growing faster compared to simulations of non-discretized space (cf. Figure 8.10b), which indicates a reduced signaling strength due to pheromone diffusion.

The presented spatial extension of the model describes particle diffusion and excluded volume effects and shows that although ML-Rules has been developed without explicit notions of space, the language concept is sufficiently expressive to describe diverse *discrete* spatial dynamics in an ad hoc way. Even the description of dynamic processes in *continuous* space seems to be possible due to the ability to specify arbitrary constraints and to assign attributes with infinitely many concrete values. However, approaches, which are aimed at spatial rule-based modeling, explicitly deal with such problems and may thus allow for modeling such systems more straightforwardly and succinctly (see, e.g., Bittig and Uhrmacher, 2010; Bittig et al., 2011).

Chapter 9

Conclusion

In this thesis, a novel rule-based language concept was proposed to facilitate multilevel modeling in systems biology. Therefore, as a starting point, various existing formal modeling approaches were investigated for their suitability to describing biological models at multiple interrelated levels of organization. Hereby identified beneficial modeling concepts and methods were then taken as a basis to develop ML-Rules, an accessible general-purpose modeling approach for describing multilevel models of biological systems.

Language features and general modeling concepts that were shown to be beneficial for this purpose are as multifaceted as the different aspects of biological multilevel models are. First of all, reaction-centric modeling approaches were shown to be generally well-suited for modeling biochemical systems, but also for describing dynamics at other organizational levels, as the reactions metaphor can be considered as an appropriate paradigm for state transitions in general. Rule-based approaches employ such a reaction-centric modeling paradigm and, moreover, they allow for effectively reducing the size of models due to rule schemata. Therefore, ML-Rules was designed as a rule-based modeling language.

Obviously substantial for the accessibility of multilevel modeling is a support for explicitly nested model structures, allowing for straightforwardly representing the hierarchical organization of biological systems, where different entities may be both vertically as well as horizontally separated from each other. However, supporting nestedness is only half way there. To appropriately reflect the variability of biological structures, i.e., their potential for dynamic rearrangements, another critical feature is to also support describing dynamically changing structures, which was realized in ML-Rules by allowing for differently nested species on the reactants and products sides of a rule and by binding unspecified content to variables that can be freely reused when specifying the rule's products, e.g., to describe migration or merging of compartments. It was also shown that by applying a decomposition function on bound solutions, the description of cell division or budding becomes possible. In any case, the nested structure of model entities describes merely their arrangement rather their behavior, for which additional constructs are needed.

For describing dynamic behavior at different organizational levels, it was shown that in the first instance it is important to allow for representing states and dynamics at any hierarchy level, so that not only atomic entities may have a state and behavior, but also the containing structures. Therefore, species in ML-Rules may have assigned arbitrarily many attributes, no matter whether they contain other species or not. These attributes are neither restricted to certain numerical values nor to any predefined (finite) value set, thereby facilitating a natural representation of rather diverse states at different levels, as was exemplified by two case studies in which the volumes of different compartments were described explicitly, for example.

By applying rule schemata and assigning new values to respective species, such high-level states may be altered dynamically, similar to any other dynamic process in ML-Rules. Thereby, with the help of arbitrary expressions, rates of rule applications can be flexibly specified and constrained, which was shown to be essential for (i) making various behavioral abstractions, (ii) appropriately representing the diversity of dynamic processes at different organizational levels, as well as (iii) modeling of interlevel causation between those. The latter was realized in ML-Rules by accessing contextual information either from a species' environment (downward causation) or its enclosed content (upward causation), for which the reaction-centric modeling paradigm proved its effectiveness again, as taking such side-effects into account can be easily achieved by expanding the set of reactants. Last but not least, it was shown by means of case studies that attributes and flexibly constrained reaction rates are sufficiently expressive to describe simple discretized spatial dynamics beyond compartmentalization, which often play an important role in multilevel modeling of biological systems, e.g., particle diffusion or other spatially constrained processes at the level of cellular tissues.

Taken all together, the proposed modeling concept can be considered as a significant step toward accessible multilevel modeling in systems biology, since it denotes a succinct and user-friendly general-purpose modeling approach addressing a wide range of different aspects that were shown to be relevant for describing such kinds of models. With *Coloured Stochastic Multilevel Multiset Rewriting* (CSMMR), Oury and Plotkin (2011) have proposed a formal language sharing many similarities with ML-Rules, e.g., rule schemata applied to nested and attributed model entities, supporting variable structures, and rate functions depending on attributes. However, CSMMR does not support links between distinct entities and, more importantly, rates rely on the law of mass action and may merely depend on attributes of reactants, but not on properties of the multiset the rule is applied to, such as the total amount of some of its constituents. Hence, describing certain behavioral abstractions and particularly upward causation may be significantly hampered. Although some of these issues have already been identified by the authors as useful extensions, it remains an open question whether they can be seamlessly integrated into their formalism. Howsoever, the approach presented in this thesis leaves some space for future research as well.

First of all, having defined a formal semantics of ML-Rules would allow for studying its expressiveness in a formal manner by proving certain theorems, e.g., that ML-Rules can encode other languages – like the π-calculus, React(C), Bigraphs, or CSMMR – or vice versa, i.e., that ML-Rules can be mapped to them, which may then lead to additional facilities for model analysis, for instance.

From the modeling point of view, a type system for attributes might further enhance the accessibility by reducing the chance to assign "incorrect" values. Different types of attributes could also help introducing an explicit notion of linkage between species, like it is supported by BNGL, for instance. Thereby, bindings may become more obvious to the modeler and explicit links may also reduce the user's responsibility for keeping things biologically relevant.

To support describing more sophisticated spatial dynamics, e.g., to model complex spatial processes at the tissue level, it would be also interesting to investigate whether some ideas of MGS (Giavitto and Michel, 2001; Michel et al., 2009) could be integrated into the modeling concept presented in this work. MGS combines rule-based modeling with so called *topological collections*, to specify which and how model entities may interact with each other. Different topological collections across various spatial scales can thereby define different local relationships of individual entities, e.g., ordinary multisets (compartments) and Delaunay triangulation.

Another useful extension of ML-Rules would be dedicated to supporting delays and general distributions, since biological phenomena – especially at higher (abstraction) levels – are not necessarily exponentially distributed. For example, cell cycle dynamics may follow a Gaussian distribution (Walker et al., 2004) and the numerous sequential reaction steps of a gene transcription and translation process may be approximated best by a simple temporal delay (Lewis, 2003). The current simulator of ML-Rules proceeds basically as a discrete event simulator, making an integration of non-exponential waiting times relatively easy in principle. However, non-Markovian processes are not *memoryless*, which leads to the problem of tracking the time consumed by each individual reaction (Mura et al., 2009) and thus might require substantial adaptations of its implementation.

Future research needs also to be done for speeding up simulations. Although the JAMES II framework – within which ML-Rules was realized – offers a coarse-grained parallel execution of multiple simulation runs (Himmelspach et al., 2008), due to the expressiveness of ML-Rules, already the single run execution of a more complex model – like the case studies presented in this thesis – takes a fairly large amount of time in the current implementation. However, the following diverse methods may help to make simulation of ML-Rules models significantly faster.

Since simulation algorithms in JAMES II are designed in a plug-in-based manner and thus not in terms of monolithic blocks, alternative sub-algorithms can be easily exploited and combined. It has been shown that the performance and suitability of algorithms depend to a large degree on the concrete model, that details (of sub-algorithms) matter, and that a suitable configuration can significantly speed up the simulation (Jeschke et al., 2011). In combination with methods that help to automatically select and configure simulators on demand (Ewald et al., 2010; Ewald, 2010), this type of simulation design supports a high flexibility for executing multilevel models efficiently. Another approach for speeding up the simulation would be to avoid the time-consuming instantiation of all possible reactions by exploiting an alternative kinetic Monte Carlo simulation approach (Danos et al., 2007b; Colvin et al., 2009; Sneddon et al., 2011) based on individual particles rather than populations of identical species. It might be also worth to explore the potentials of a fine-grained parallel execution. The challenge here is to reduce the synchronization overhead between different computational nodes as much as possible, e.g., by partitioning the system appropriately, for which the nested structure of a multilevel model might denote a suitable indicator, as *"the relations inside a subsystem are [typically] stronger than the interrelations between subsystems"* (Timpf, 1999, p. 129). Lastly, since multilevel models often operate at different temporal scales, the calculation efforts for simulating such models in a pure discrete event manner might – despite the above optimizations – easily become prohibitive. Therefore, hybrid simulation approaches (Takahashi et al., 2004; MacNamara and Burrage, 2010; Coppo et al., 2010a) could be exploited in the future, which allow certain processes involving abundant species to be interpreted continuously by numerically solving according ODEs.

To get back to the actual topic of this thesis, it was shown once again that the syntax of a language strongly influences its accessibility (Henzinger et al., 2009), i.e., the syntax is that part of a language that really matters. This was shown by minor syntactical improvements of the basic ML-Rules notation, however, it becomes much more obvious when comparing the proposed multilevel modeling concept with diverse other modeling approaches discussed in this thesis, most of them employing the same basic mathematical CTMC semantics like ML-Rules, e.g., stochastic Petri nets, the stochastic π-calculus, BioAmbients, and Stochastic Bigraphs – to name just a few. Although it should theoretically be possible to describe the same dynamics with each of these languages, it was shown that in practice this is not the case. Hence, for the transformation process from an informal mental model to a formal model representation, it is less the quest to find a formalism that is *somehow* able to express certain phenomena, but one with which this can be done *easily.*

Appendix A

Concrete Syntax of ML-Rules

A.1   EBNF Grammar (W3C notation)

```
definitions
        ::= constants? species_definitions initial_solution rules EOF
constants
        ::= NAME ':' expression ';' constants?
species_definitions
        ::= species_def+
species_def
        ::= NAME '(' INT? ')' ';'
initial_solution
        ::= '>>INIT' '[' (init_element ( '+' init_element )*)? ']' ';'
init_element
        ::= expression NAME '(' assignments ')' ( '[' init_element ( '+' init_element )* ']' )?
        |  NAME '(' assignments ')' ( '[' init_element ( '+' init_element )* ']' )?
        |  expression NAME ( '(' ')' )? '[' init_element ( '+' init_element )* ']'
        |  NAME ( '(' ')' )? '[' init_element ( '+' init_element )* ']'
        |  NAME ( '(' ')' )?
        |  expression NAME ( '(' ')' )?
        |  for_loop_init
assignments
        ::= expression ( ',' assignments )?
rules  ::= rule+
rule   ::= reactants '->' products? '@' expression ';'
reactants
        ::= NAME '?'
        |  reactant ( '+' reactants )?
reactant
        ::= NAME '(' pattern_matching ')' '[' reactants ']' ( ':' NAME )?
        |  NAME ( '(' ')' )? '[' reactants ']' ( ':' NAME )?
        |  NAME '(' pattern_matching ')' ( ':' NAME )?
        |  NAME ( '(' ')' )? ( ':' NAME )?
        |  expression NAME '(' pattern_matching ')' '[' reactants ']' ( ':' NAME )?
```

```
        | expression NAME ( '(' ')' )? '[' reactants ']' ( ':' NAME )?
        | expression NAME '(' pattern_matching ')' ( ':' NAME )?
        | expression NAME ( '(' ')' )? ( ':' NAME )?
pattern_matching
        ::= expression ( ',' pattern_matching )?
products
        ::= product ( '+' products )?
product
        ::= expression NAME '?'
        | NAME '?'
        | NAME '(' assignments ')' '[' products ']'
        | NAME ( '(' ')' )? '[' products ']'
        | NAME '(' assignments ')'
        | NAME ( '(' ')' )?
        | expression NAME '(' assignments ')' '[' products ']'
        | expression NAME ( '(' ')' )? '[' products ']'
        | expression NAME '(' assignments ')'
        | expression NAME ( '(' ')' )?
expressions
        ::= expression ( ',' expressions )?
expression
        ::= term ( ( '+' | '-' ) term )*
        | NU
term    ::= term_power ( ( '*' | '/' ) term_power )*
term_power
        ::= unary_op ( '^' unary_op )*
unary_op
        ::= factor
        | '-' factor
factor ::= INT
        | FLOAT
        | STRING
        | COUNT
        | function
        | NAME
        | boolean_expression
        | '(' expression ')'
        | 'if' boolean_expression 'then' expression 'else' expression
function
        ::= NAME '(' expressions? ')'
boolean_expression
        ::= andexpression ( '||' andexpression )*
andexpression
        ::= notexpression ( '&&' notexpression )*
notexpression
        ::= '!' atom
        | atom
atom    ::= condition
        | '(' boolean_expression ')'
condition
        ::= '(' expression '<' expression ')'
```

```
                 | '(' expression '>' expression ')'
                 | '(' expression '>=' expression ')'
                 | '(' expression '<=' expression ')'
                 | '(' expression '==' expression ')'
                 | '(' expression '!=' expression ')'
                 | 'true'
                 | 'false'
                 | NAME
for_loop_init
        ::= 'for' NAME ':' expression 'while' boolean_expression 'with' expression
            for_loop_init_body ';'?
for_loop_init_body
        ::= init_element
          | '[' init_element ( '+' init_element )* ']'
COUNT   ::= '#' NAME
NU      ::= '$' NAME
NAME    ::= ( [a-z] | [A-Z] | '_' ) ( [a-z] | [A-Z] | [0-9] | '_' )*
INT     ::= [0-9]+
FLOAT   ::= INT
          | [0-9]+ '.' [0-9]* EXPONENT?
          | '.' [0-9]+ EXPONENT?
          | [0-9]+ EXPONENT
EXPONENT
        ::= ( 'e' | 'E' ) ( '+' | '-' )? [0-9]+
STRING ::= '"' ( ESC_SEQ | [^\'] )* '"'
HEX_DIGIT
        ::= [0-9]
          | [a-f]
          | [A-F]
ESC_SEQ
        ::= '\' ( 'b' | 't' | 'n' | 'f' | 'r' | '"' | '\'' | '\' )
          | UNICODE_ESC
          | OCTAL_ESC
OCTAL_ESC
        ::= '\' [0-3] [0-7] [0-7]
          | '\' [0-7] [0-7]
          | '\' [0-7]
UNICODE_ESC
        ::= '\' 'u' HEX_DIGIT HEX_DIGIT HEX_DIGIT HEX_DIGIT
IGNORED
        ::= COMMENT
          | WS
COMMENT?
        ::= '//' [^#xA#xD]* ( #xD? #xA )?
          | '/*' .* '*/'
WS      ::= ' '
          | #x9
          | #xD
          | #xA
EOF     ::= $
```

221

Appendix B

Yeast Cell Proliferation Models Encoded in the Concrete Textual Syntax

B.1   Cell Cycle Control by Both Downward and Upward Causation

```
// parameters
cdc2tot:  1000;
Td:       116;
k1:       0.015*cdc2tot;
k2:       200;
k3:       180;
k3prime:  0.018;
k4:       4.5;
k5:       0.6;
k6:       1.0;
k7:       1e6;
k8:       1e6;
k9:       1e6;
t7:       250;
t8:       70;
t9:       20;

// species definitions
Cell(2);
Cdc2();
Cyclin();
CyclinP();
MPFi();
MPFa();

// initial solution
>>INIT[ Cell(1,'G1')[cdc2tot Cdc2] ];

// rule schemata
Cell(v,p)[s?]           -> Cell(v,p)[Cyclin + s?]    @ k1;
```

```
Cyclin:y + Cdc2:d        -> MPFi                    @ k2*#y*#d;
MPFi:m                   -> MPFa                    @ k3prime*#m;
MPFa:a + MPFi:m          -> 2 MPFa                  @ k3*((#a/cdc2tot)^2)*#m;
Cell(v,p)[MPFa:m + s?]   -> Cell(v,p)[CyclinP + Cdc2 + s?]  @ (k4/v)*#m;
CyclinP:y                ->                         @ k5*#y;
Cell(v,p)[s?]            -> Cell(v+(1/Td),p)[s?]    @ if (p=='G1') || (p=='SG2') then k6 else 0;
Cell(v,'G1')[MPFi:m + s?] -> Cell(v,'SG2')[MPFi + s?]  @ if (#m>t7) then k7 else 0;
Cell(v,'SG2')[MPFa:m + s?] -> Cell(v,'M')[MPFa + s?]  @ if (#m>t8) then k8 else 0;
Cell(v,'M')[MPFa:m + s?]  -> Cell(v/2,'G1')[MPFa + s?] @ if (#m<t9) then k9 else 0;
```

B.2 Multicellular Proliferation and Mating Type Switching

```
// parameters
cdc2tot:  1000;
Td:       116;
k1:       0.015*cdc2tot;
k2:       200;
k3:       180;
k3prime:  0.018;
k4:       4.5;
k5:       0.6;
k6:       1.0;
k7:       1e6;
k8:       1e6;
k9:       1e6;
t7:       250;
t8:       70;
t9:       20;
kdeath:   0.006;

// species definitions
Cell(4);
Cdc2();
Cyclin();
CyclinP();
MPFi();
MPFa();

// initial solution
>>INIT[ 100 Cell(1,'G1','P','U')[cdc2tot Cdc2] ];

// rule schemata
Cell(v,p,t,w)[s?]:c -> Cell(v,p,t,w)[Cyclin + s?] @ k1*#c;

Cyclin:y + Cdc2:d -> MPFi @ k2*#y*#d;

MPFi:m -> MPFa @ k3prime*#m;

MPFa:a + MPFi:m -> 2 MPFa @ k3*((#a/cdc2tot)^2)*#m;

CyclinP:y -> @ k5*#y;

Cell(v,p,t,w)[s?]:c -> Cell(v+(1/Td),p,t,w)[s?] @ if (p=='G1') || (p=='SG2') then k6*#c else 0;

Cell(v,p,t,w)[s?]:c -> @ kdeath*#c;

Cell(v,p,t,w)[MPFa:m + s?]:c -> Cell(v,p,t,w)[CyclinP + Cdc2 + s?] @ (k4/v)*#m*#c;

Cell(v,'G1',t,w)[MPFi:m + s?]:c -> Cell(v,'SG2',t,w)[MPFi + s?] @ if (#m>t7) then k7*#c else 0;

Cell(v,'SG2',t,w)[MPFa:m + s?]:c -> Cell(v,'M',t,w)[MPFa + s?] @ if (#m>t8) then k8*#c else 0;
```

```
Cell(v,'M',t,'U')[MPFa:m + s?]:c -> Cell(v/2,'G1',t,'S')[MPFa + s?] + Cell(v/2,'G1',t,'U')[MPFa + s?] @
    if (#m<t9) then k9*#c else 0;

Cell(v,'M',t,'S')[MPFa:m + s?]:c -> Cell(v/2,'G1',t,'S')[MPFa + s?] + Cell(v/2,'G1',if (t=='P') then
    'M' else 'P','U')[MPFa + s?] @ if (#m<t9) then k9*#c else 0;
```

B.3 Pheromone-dependent Cell Cycle Dynamics in a Single Cell

```
// parameters
cdc2tot: 1000;
Td:      232;
k1:      0.015*cdc2tot;
k2:      200;
k3:      180;
k3prime: 0.018;
k4:      4.5;
k5:      0.6;
k6:      1.0;
k7:      1e6;
k8:      1e6;
k9:      1e6;
t7:      250;
t8:      70;
t9:      20;
k10:     1.0;
k11:     1.5;
K11:     800;
k12:     0.02;
PheInit: 0;      // set to 200/600 for studying effect of pheromone response

// species definitions
Cell(2);
Cdc2();
Cyclin();
CyclinP();
MPFi();
MPFa();
MPFr();
Phe();

// initial solution
>>INIT[ PheInit Phe + Cell(1,'G1')[cdc2tot Cdc2] ];

// rule schemata
Cell(v,p)[s?] -> Cell(v,p)[Cyclin + s?] @ k1;

Cyclin:y + Cdc2:d -> MPFi @ k2*#y*#d;

MPFi:m -> MPFa @ k3prime*#m;

MPFa:a + MPFi:m -> 2 MPFa @ k3*((#a/cdc2tot)^2)*#m;

Cell(v,p)[MPFa:m + s?] -> Cell(v,p)[CyclinP + Cdc2 + s?] @ (k4/v)*#m;

CyclinP:y -> @ k5*#y;

Cell(v,p)[s?] -> Cell(v+(1/Td),p)[s?] @ if (p=='G1') || (p=='SG2') then k6 else 0;

Cell(v,'G1')[MPFi:m + s?] -> Cell(v,'SG2')[MPFi + s?] @ if (#m>t7) then k7 else 0;
```

```
Cell(v,'SG2')[MPFa:m + s?] -> Cell(v,'M')[MPFa + s?] @ if (#m>t8) then k8 else 0;

Cell(v,'M')[MPFa:m + s?] -> Cell(v/2,'G1')[MPFa + s?] @ if (#m<t9) then k9 else 0;

Phe:p + Cell(v,ph)[MPFi:m + s?]:c -> Phe + Cell(v,ph)[MPFr + s?] @
    (((k11*(#p^3))/((K11^3)+(#p^3)))*(1/(v^2)))*#m*#c;

MPFr:m -> MPFi @ k12*#m;
```

B.4 Multicellular Model of Yeast Cell Proliferation and Intercellular Communication

```
// parameters
cdc2tot: 1000;
Td:      232;
k1:      0.015*cdc2tot;
k2:      200;
k3:      180;
k3prime: 0.018;
k4:      4.5;
k5:      0.6;
k6:      1.0;
k7:      1e6;
t7:      250;
k8:      1e6;
t8:      70;
k9:      1e6;
t9:      20;
k10:     1.0; // set to 0 for turning off pheromone secretion
k11:     1.5;
K11:     800;
k12:     0.02;
k13:     0.005;
k14:     1.0;
k15:     1e-4;

// species
Cell(4);
Cdc2();
Cyclin();
CyclinP();
MPFi();
MPFa();
MPFr();
Phe(1);
Sxa2();

// initial solution
>>INIT[ Cell(1,'G1','P','U')[cdc2tot Cdc2] + Cell(1,'G1','M','U')[cdc2tot Cdc2] ];

// rule schemata
Cell(v,p,t,w)[s?]:c -> Cell(v,p,t,w)[Cyclin + s?] @ k1*#c;

Cyclin:y + Cdc2:d -> MPFi @ k2*#y*#d;

MPFi:m -> MPFa @ k3prime*#m;

MPFa:a + MPFi:m -> 2 MPFa @ k3*((#a/cdc2tot)^2)*#m;

Cell(v,p,t,w)[MPFa:m + s?]:c -> Cell(v,p,t,w)[CyclinP + Cdc2 + s?] @ (k4/v)*#m*#c;

CyclinP:y -> @ k5*#y;
```

```
Cell(v,p,t,w)[s?]:c -> Cell(v+(1/Td),p,t,w)[s?] @ if (p=='G1') || (p=='SG2') then k6**c else 0;

Cell(v,'G1',t,w)[MPFi:m + s?]:c -> Cell(v,'SG2',t,w)[MPFi + s?] @ if (#m>t7) then k7**c else 0;

Cell(v,'SG2',t,w)[MPFa:m + s?]:c -> Cell(v,'M',t,w)[MPFa + s?] @ if (#m>t8) then k8**c else 0;

Cell(v,'M',t,'U')[MPFa:m + sc?]:c -> Cell(v/2,'G1',t,'S')[MPFa + sc?] + Cell(v/2,'G1',t,'U')[MPFa +
    sc?] @ if (#m<t9) then k9**c else 0;

Cell(v,'M',t,'S')[MPFa:m + sc?]:c -> Cell(v/2,'G1',t,'S')[MPFa + sc?] + Cell(v/2,'G1',if (t=='P') then
    'M' else 'P','U')[MPFa + sc?] @ if (#m<t9) then k9**c else 0;

Cell(v,p,t,w)[s?]:c -> Cell(v,p,t,w)[s?] + Phe(t) @ k10**c;

Cell(v,p,'M',w)[s?]:c -> Cell(v,p,'M',w)[s?] + Sxa2 @ 0.1*k10**c;

Sxa2:x -> @ k13**x;

Sxa2:x + Phe('P'):p -> Sxa2 @ k15**x**p;

Phe(tp):p + Cell(v,p,tc,w)[MPFi:m + s?]:c -> Phe(tp) + Cell(v,p,tc,w)[MPFr + s?] @ if (tp!=tc) then
    (((k11*(#p^3))/((K11^3)+(#p^3)))*(1/(v^2)))**m**c else 0;

MPFr:m -> MPFi @ k12**m;

Phe(t):p -> @ k13**p;
```

B.5 Proliferating and Communicating Cells in Space

```
// parameters
cdc2tot: 1000;
Td:      232;
k1:      0.015*cdc2tot;
k2:      200;
k3:      180;
k3prime: 0.018;
k4:      4.5;
k5:      0.6;
k6:      1.0;
k7:      1e6;
t7:      250;
k8:      1e6;
t8:      70;
k9:      1e6;
t9:      20;
k10:     1.0;
k11:     1.5;
K11:     800;
k12:     0.02;
k13:     0.005;
k14:     1.0;
k15:     1e-4;
xmax:    5;
ymax:    5;

// species
Cell(4);
Cdc2();
Cyclin();
CyclinP();
```

```
MPFi();
MPFa();
MPFr();
Phe(1);
Sxa2();
SV(3);

// initial solution
>>INIT[
 for y:1 while (y<=2) with y+1 [
  for x:1 while (x<=xmax) with x+1 [SV(x,y,0)]
 ]
 +
 for x:1 while (x<=2) with x+1 [SV(x,3,0)]
 +
 SV(3,3,2)[
  Cell(1,'G1','P','U')[cdc2tot Cdc2] +
  Cell(1,'G1','M','U')[cdc2tot Cdc2]
 ]
 +
 for x:4 while (x<=xmax) with x+1 [SV(x,3,0)]
 +
 for y:4 while (y<=ymax) with y+1 [
  for x:1 while (x<=xmax) with x+1 [SV(x,y,0)]
 ]
];

// rule schemata
Cell(v,p,t,w)[s?]:c -> Cell(v,p,t,w)[Cyclin + s?] @ k1*#c;

Cyclin:y + Cdc2:d -> MPFi @ k2*#y*#d;

MPFi:m -> MPFa @ k3prime*#m;

MPFa:a + MPFi:m -> 2 MPFa @ k3*((#a/cdc2tot)^2)*#m;

Cell(v,p,t,w)[MPFa:m + s?]:c -> Cell(v,p,t,w)[CyclinP + Cdc2 + s?] @ (k4/v)*#m*#c;

CyclinP:y -> @ k5*#y;

Cell(v,p,t,w)[s?]:c -> Cell(v+(1/Td),p,t,w)[s?] @ if (p=='G1') || (p=='SG2') then k6*#c else 0;

Cell(v,'G1',t,w)[MPFi:m + s?]:c -> Cell(v,'SG2',t,w)[MPFi + s?] @ if (#m>t7) then k7*#c else 0;

Cell(v,'SG2',t,w)[MPFa:m + s?]:c -> Cell(v,'M',t,w)[MPFa + s?] @ if (#m>t8) then k8*#c else 0;

SV(x,y,n)[Cell(v,'M',t,'U')[MPFa:m + sc?]:c + sv?] -> SV(x,y,n+1)[Cell(v/2,'G1',t,'S')[MPFa + sc?] +
     Cell(v/2,'G1',t,'U')[MPFa + sc?] + sv?] @ if (#m<t9) then k9*#c else 0;

SV(x,y,n)[Cell(v,'M',t,'S')[MPFa:m + sc?]:c + sv?] -> SV(x,y,n+1)[Cell(v/2,'G1',t,'S')[MPFa + sc?] +
     Cell(v/2,'G1',if (t=='P') then 'M' else 'P','U')[MPFa + sc?] + sv?] @ if (#m<t9) then k9*#c else
     0;

Cell(v,p,t,w)[s?]:c -> Cell(v,p,t,w)[s?] + Phe(t) @ k10*#c;

Cell(v,p,'M',w)[s?]:c -> Cell(v,p,'M',w)[s?] + Sxa2 @ 0.1*k10*#c;

Sxa2:x -> @ k13*#x;

Sxa2:x + Phe('P'):p -> Sxa2 @ k15*#x*#p;

Phe(tp):p + Cell(v,p,tc,w)[MPFi:m + s?]:c -> Phe(tp) + Cell(v,p,tc,w)[MPFr + s?] @ if (tp!=tc) then
     (((k11*(#p^3))/((k11^3)+(#p^3)))*(1/(v^2)))*#m*#c else 0;

MPFr:m -> MPFi @ k12*#m;

SV(xi,yi,ni)[Phe(t):p + si?] + SV(xj,yj,nj)[sj?] -> SV(xi,yi,ni)[si?] + SV(xj,yj,nj)[Phe(t) + sj?] @ if
     (((xi==xj) && ((yi==yj+1) || (yi==yj-1))) || ((yi==yj) && ((xi==xj+1) || (xi==xj-1)))) then
     (k13/4)*#p else 0;
```

```
SV(x,y,n)[Phe(t):p + s?] -> SV(x,y,n)[s?] @ if (x==1) || (x==xmax) then (k13/4)*#p else 0;

SV(x,y,n)[Phe(t):p + s?] -> SV(x,y,n)[s?] @ if (y==1) || (y==ymax) then (k13/4)*#p else 0;

SV(xi,yi,ni)[Cell(v,p,t,w)[sc?] + si?] + SV(xj,yj,nj)[sj?] -> SV(xi,yi,ni-1)[si?] +
    SV(xj,yj,nj+1)[Cell(v,p,t,w)[sc?] + sj?] @ if (((xi==xj) && ((yi==yj+1) || (yi==yj-1))) ||
    ((yi==yj) && ((xi==xj+1) || (xi==xj-1)))) && (ni>4) then k14*((ni^2)/(1+(nj^2))) else 0;
```

Bibliography

Adra, S., Sun, T., MacNeil, S., Holcombe, M., and Smallwood, R. (2010). Development of a Three Dimensional Multiscale Computational Model of the Human Epidermis. *PLoS ONE*, 5(1):e8511.

Alves, R., Antunes, F., and Salvador, A. (2006). Tools for kinetic modeling of biochemical networks. *Nature Biotechnology*, 24(6):667–672.

Amir-Kroll, H., Sadot, A., Cohen, I., and Harel, D. (2008). GemCell: A generic platform for modeling multi-cellular biological systems. *Theoretical Computer Science*, 391(3):276–290.

Anderson, A. R. A. (2005). A hybrid mathematical model of solid tumour invasion: the importance of cell adhesion. *Mathematical Medicine and Biology*, 22(2):163–186.

Anderson, P. W. (1972). More is Different: Broken symmetry and the nature of the hierarchical structure of science. *Science*, 177(4047):393–396.

Angermann, B. R., Klauschen, F., Garcia, A. D., Prustel, T., Zhang, F., Germain, R. N., and Meier-Schellersheim, M. (2012). Computational modeling of cellular signaling processes embedded into dynamic spatial contexts. *Nature Methods*, 9(3):283–289.

Ardelean, I. I. and Cavaliere, M. (2003). Modelling biological processes by using a probabilistic P system software. *Natural Computing*, 2(2):173–197.

Artavanis-Tsakonas, S., Rand, M. D., and Lake, R. J. (1999). Notch Signaling: Cell Fate Control and Signal Integration in Development. *Science*, 284(5415):770–776.

Baader, F., Calvanese, D., McGuinness, D., Nardi, D., and Patel-Schneider, P., editors (2003). *The Description Logic Handbook: Theory, Implementation and Applications.* Cambridge University Press.

Baeten, J. (2005). A Brief History of Process Algebra. *Theoretical Computer Science*, 335(2–3):131–146.

Balci, O. (1988). The implementation of four conceptual frameworks for simulation modeling in high-level languages. In Abrams, M., Haigh, P., and Comfort, J., editors, *Proceedings of the 1988 Winter Simulation Conference*, pages 287–295.

Band, L. R. and King, J. R. (2012). Multiscale modelling of auxin transport in the plant-root elongation zone. *Journal of Mathematical Biology*, 65:743–785.

Barberis, M., Klipp, E., Vanoni, M., and Alberghina, L. (2007). Cell Size at S Phase Initiation: An Emergent Property of the G_1/S Network. *PLoS Computational Biology*, 3(4):e64.

Barbuti, R., Maggiolo-Schettini, A., Milazzo, P., and Troina, A. (2006). A Calculus of Looping Sequences for Modelling Microbiological Systems. *Fundamenta Informaticae*, 72(1–3):21–35.

Barbuti, R., Maggiolo–Schettini, A., Milazzo, P., and Troina, A. (2007). The Calculus of Looping Sequences for Modeling Biological Membranes. In Eleftherakis, G., Kefalas, P., Păun, G., Rozenberg, G., and Salomaa, A., editors, *Membrane Computing*, volume 4860 of *Lecture Notes in Computer Science*, pages 54–76. Springer.

Baron, M., Aslam, H., Flasza, M., Fostier, M., Higgs, J., Mazaleyrat, S., and Wilkin, M. (2002). Multiple levels of Notch signal regulation (Review). *Molecular Membrane Biology*, 19(1):27–38.

Beach, D. (1983). Cell type switching by DNA transposition in fission yeast. *Nature*, 305:682–688.

Beach, D. H. and Klar, A. J. (1984). Rearrangements of the transposable mating-type cassettes of fission yeast. *The EMBO Journal*, 3(3):603–610.

Bedau, M. A. (1997). Weak Emergence. *Noûs*, 31:375–399.

Benjamin, P., Erraguntla, M., Delen, D., and Mayer, R. (1998). Simulation modeling at multiple levels of abstraction. In Medeiros, D., Watson, E., Carson, J., and Manivannan, M., editors, *Proceedings of the 1998 Winter Simulation Conference*, pages 391–398.

Berg, J. M., Tymoczko, J. L., and Stryer, L. (2002). *Biochemistry*. W. H. Freeman, 5th edition.

Bhalla, U. S. (2004). Signaling in Small Subcellular Volumes. I. Stochastic and Diffusion Effects on Individual Pathways. *Biophysical Journal*, 87(2):733–744.

Bittig, A. T., Haack, F., Maus, C., and Uhrmacher, A. M. (2011). Adapting rule-based model descriptions for simulating in continuous and hybrid space. In *Proceedings of the 9th International Conference on Computational Methods in Systems Biology*, pages 161–170.

Bittig, A. T. and Uhrmacher, A. M. (2010). Spatial Modeling in Cell Biology at Multiple Levels. In Johansson, B., Jain, S., Montoya-Torres, J., Hugan, J., and Yücesan, E., editors, *Proceedings of the 2010 Winter Simulation Conference*, pages 608–619.

Blinov, M. L. and Moraru, I. I. (2012). Leveraging Modeling Approaches: Reaction Networks and Rules. In Goryanin, I. I. and Goryachev, A. B., editors, *Advances in Systems Biology*, volume 736 of *Advances in Experimental Medicine and Biology*, pages 517–530. Springer.

Bocharov, G. A. and Rihan, F. A. (2000). Numerical modelling in biosciences using delay differential equations. *Journal of Computational and Applied Mathematics*, 125(1–2):183–199.

Borland, S. and Vangheluwe, H. (2003). Transforming Statecharts to DEVS. In *Summer Computer Simulation Conference (Student Workshop)*, pages S154–S159.

Bradley, D. F. (1968). Multilevel Systems and Biology–View of a Submolecular Biologist. In Mesarović, M. D., editor, *Systems Theory and Biology: Proceedings of the III Systems Symposium at Case Institute of Technology*, pages 38–58. Springer, Berlin, Heidelberg, New York.

Bray, S. J. (2006). Notch signalling: a simple pathway becomes complex. *Nature Reviews Molecular Cell Biology*, 7(9):678–689.

Brodo, L., Degano, P., and Priami, C. (2007). A Stochastic Semantics for BioAmbients. In Malyshkin, V., editor, *Parallel Computing Technologies*, volume 4671 of *Lecture Notes in Computer Science*, pages 22–34. Springer.

Brook, B. and Waters, S. (2008). Mathematical challenges in integrative physiology. *Journal of Mathematical Biology*, 56(6):893–896.

Bunge, M. (1959). *Causality: The Place of the Causal Principle in Modern Science*. Harvard University Press.

Bunge, M. (1969). The Metaphysics, Epistemology and Methodology of Levels. In Whyte, L., Wilson, A., and Wilson, D., editors, *Hierarchical structures*, pages 17–28. Elsevier, New York.

Bunge, M. (1977). General Systems and Holism. In *General Systems: Yearbook of the Society for the Advancement of General Systems Theory*, volume 22, pages 87–90.

Busi, N. (1999). Mobile Nets. In Ciancarini, P., Fantechi, A., and Gorrieri, R., editors, *Formal Methods for Open Object-Based Distributed Systems*, pages 51–66. Kluwer Academic Publishers.

Butcher, E. C., Berg, E. L., and Kunkel, E. J. (2004). Systems biology in drug discovery. *Nature Biotechnology*, 22(10):1253–1259.

Calder, M., Duguid, A., Gilmore, S., and Hillston, J. (2006a). Stronger Computational Modelling of Signalling Pathways Using Both Continuous and Discrete-State Methods. In Priami, C., editor, *Computational Methods in Systems Biology*, volume 4210 of *Lecture Notes in Computer Science*, pages 63–77. Springer.

Calder, M., Vyshemirsky, V., Gilbert, D., and Orton, R. (2006b). Analysis of Signalling Pathways Using Continuous Time Markov Chains. In Priami, C. and Plotkin, G., editors, *Transactions on Computational Systems Biology VI*, volume 4220 of *Lecture Notes in Computer Science*, pages 44–67. Springer.

Campbell, D. T. (1974). 'Downward Causation' in Hierarchically Organised Biological Systems. In Ayala, F. J. and Dobzhansky, T., editors, *Studies in the Philosophy of Biology: Reduction and Related Problems*, pages 179–186. Macmillan Press.

Cardelli, L. (2005). Brane Calculi: Interactions of Biological Membranes. In Danos, V. and Schachter, V., editors, *Computational Methods in Systems Biology*, volume 3082 of *Lecture Notes in Computer Science*, pages 257–278. Springer.

Cardelli, L. (2007). A Process Algebra Master Equation. In *Proceedings of the Fourth International Conference on the Quantitative Evaluation of Systems*, pages 219–226.

Cardelli, L. and Gordon, A. (1998). Mobile ambients. In Nivat, M., editor, *Foundations of Software Science and Computation Structures*, volume 1378 of *Lecture Notes in Computer Science*, pages 140–155. Springer.

Casini, L. (2011). Interview with Olaf Wolkenhauer. *The Reasoner*, 5(9):139–143.

Cellier, F. E. (1991). *Continuous System Modeling*. Springer.

Chabrier-Rivier, N., Fages, F., and Soliman, S. (2005). The Biochemical Abstract Machine BIOCHAM. In Danos, V. and Schachter, V., editors, *Computational Methods in Systems Biology*, volume 3082/2005 of *Lecture Notes in Computer Science*, pages 172–191. Springer.

Chalmers, D. J. (2006). Strong and Weak Emergence. In Clayton, P. and Davies, P., editors, *The Re-Emergence of Emergence: The Emergentist Hypothesis from Science to Religion*, pages 244–256. Oxford University Press.

Chaouiya, C. (2007). Petri net modelling of biological networks. *Briefings in Bioinformatics*, 8(4):210–219.

Chigurupati, S., Arumugam, T. V., Son, T. G., Lathia, J. D., Jameel, S., Mughal, M. R., Tang, S.-C., Jo, D.-G., Camandola, S., Giunta, M., Rakova, I., McDonnell, N., Miele, L., Mattson, M. P., and Poosala, S. (2007). Involvement of Notch Signaling in Wound Healing. *PLoS ONE*, 2(11):e1167.

Chow, A. and Zeigler, B. (1994). Parallel DEVS: A parallel, hierarchical, modular, modeling formalism. In Tew, J., Manivannan, S., Sadowski, D., and Seila, A., editors, *Proceedings of the 1994 Winter Simulation Conference*, pages 716–722.

Christensen, S. and Damgaard Hansen, N. (1994). Coloured Petri Nets extended with channels for synchronous communication. In Valette, R., editor, *Application and Theory of Petri Nets 1994*, volume 815 of *Lecture Notes in Computer Science*, pages 159–178. Springer.

Chwif, L., Barretto, M. R. P., and Paul, R. J. (2000). On Simulation Model Complexity. In Joines, J., Barton, R., Kang, K. ., and Fishwick, P., editors, *Proceedings of the 2000 Winter Simulation Conference*, pages 449–455.

Cickovski, T., Aras, K., Swat, M., Merks, R., Glimm, T., Hentschel, H., Alber, M., Glazier, J., Newman, S., and Izaguirre, J. (2007). From Genes to Organisms Via the Cell: A Problem-Solving Environment for Multicellular Development. *Computing in Science & Engineering*, 9(4):50–60.

Cohen, D. M. and Bergman, R. N. (1994). SYNTAX: A Rule-Based Stochastic Simulation of the Time-Varying Concentrations of Positional Isotopomers of Metabolic Intermediates. *Computers and Biomedical Research*, 27(2):130–147.

Cohen, S. D. and Hindmarsh, A. C. (1996). CVODE, a stiff/nonstiff ODE solver in C. *Computers in Physics*, 10(2):138–143.

Collier, J. R., Monk, N. A., Maini, P. K., and Lewis, J. H. (1996). Pattern Formation by Lateral Inhibition with Feedback: a Mathematical Model of Delta-Notch Intercellular Signalling. *Journal of Theoretical Biology*, 183(4):429–446.

Colvin, J., Monine, M. I., Faeder, J. R., Hlavacek, W. S., Hoff, D. D. V., and Posner, R. G. (2009). Simulation of large-scale rule-based models. *Bioinformatics*, 25(7):910–917.

Coppo, M., Damiani, F., Drocco, M., Grassi, E., Sciacca, E., Spinella, S., and Troina, A. (2010a). Hybrid Calculus of Wrapped Compartments. In G.Ciobanu and M.Koutny, editors, *Proceedings Compendium of the Fourth Workshop on Membrane Computing and Biologically Inspired Process Calculi 2010*, volume 40 of *EPTCS*, pages 102–120.

Coppo, M., Damiani, F., Drocco, M., Grassi, E., and Troina, A. (2010b). Stochastic Calculus of Wrapped Compartment. In Pierro, A. D. and Norman, G., editors, *Proceedings Eighth Workshop on Quantitative Aspects of Programming Languages*, volume 28 of *EPTCS*, pages 82–98.

Corning, P. A. (2002). The Re-emergence of "Emergence": A Venerable Concept in Search of a Theory. *Complexity*, 7(6):18–30.

Cristini, V. and Lowengrub, J. (2010). *Multiscale Modeling of Cancer: An Integrated Experimental and Mathematical Modeling Approach*. Cambridge University Press.

Curtis, B. (1980). Measurement and Experimentation in Software Engineering. *Proceedings of the IEEE*, 68(9):1144–1157.

Dada, J. O. and Mendes, P. (2011). Multi-scale modelling and simulation in systems biology. *Integrative Biology*, 3(2):86–96.

Danos, V. (2009). Agile Modelling of Cellular Signalling (Invited Paper). *Electronic Notes in Theoretical Computer Science*, 229(4):3–10.

Danos, V., Feret, J., Fontana, W., and Harmer, R. (2007a). Rule-Based Modelling of Cellular Signalling. In Caires, L. and Vasconcelos, V. T., editors, *CONCUR 2007 – Concurrency Theory: 18th International Conference*, volume 4703 of *Lecture Notes in Computer Science*, pages 17–41. Springer.

Danos, V., Feret, J., Fontana, W., Harmer, R., and Krivine, J. (2009). Rule-Based Modelling and Model Perturbation. In Priami, C., Back, R.-J., and Petre, I., editors, *Transactions on Computational Systems Biology XI*, volume 5750 of *Lecture Notes in Computer Science*, pages 116–137. Springer.

Danos, V., Feret, J., Fontana, W., and Krivine, J. (2007b). Scalable Simulation of Cellular Signaling Networks. In Shao, Z., editor, *Programming Languages and Systems*, volume 4807 of *Lecture Notes in Computer Science*, pages 139–157. Springer.

Daum, T. and Sargent, R. G. (1999). Scaling, Hierarchical Modeling, and Reuse in an Object-oriented Modeling and Simulation System. In Farrington, P., Nembhard, H., Sturrock, D., and Evans, G., editors, *Proceedings of the 1999 Winter Simulation Conference*, pages 1470–1477.

de Jong, H. (2002). Modeling and Simulation of Genetic Regulatory Systems: A Literature Review. *Journal of Computational Biology*, 9(1):67–103.

Degenring, D., Röhl, M., and Uhrmacher, A. M. (2003). Discrete Event Simulation for a Better Understanding of Metabolite Channeling – A System Theoretic Approach. In Priami, C., editor, *Computational Methods in Systems Biology*, volume 2602 of *Lecture Notes in Computer Science*, pages 114–126. Springer.

Degenring, D., Röhl, M., and Uhrmacher, A. M. (2004). Discrete event, multi-level simulation of metabolite channeling. *Biosystems*, 75(1–3):29–41.

Derosa, P. and Cagin, T., editors (2010). *Multiscale Modeling: From Atoms to Devices*. CRC Press.

Derrick, E. J., Balci, O., and Nance, R. E. (1989). A comparison of selected conceptual frameworks for simulation modeling. In MacNair, E., Musselman, K., and Heidelberger, P., editors, *Proceedings of the 1989 Winter Simulation Conference*, pages 711–718.

Dexter, N., Kruse, K., Nutaro, J., and Ward, R. (2009). A Computational Model of Cell Migration in Response to Biochemical Diffusion. In *Proceed-*

ings of the Biomedical Science and Engineering Conference at ORNL, pages 1–4. IEEE.

Dijkstra, E. W. (1975). Guarded Commands, Nondeterminacy and Formal Derivation of Programs. *Communications of the ACM*, 18(8):453–457.

E, W. (2011). *Principles of Multiscale Modeling*. Cambridge University Press.

E, W. and Engquist, B. (2003). Multiscale modeling and computation. *Notices of the AMS*, 50(9):1062–1070.

Edmonds, B. (1999). *Syntactic Measures of Complexity*. PhD thesis, University of Manchester, Manchester, UK.

Edmonds, B. (2000). Complexity and Scientific Modelling. *Foundations of Science*, 5(3):379–390.

Efroni, S., Harel, D., and Cohen, I. (2003). Toward Rigorous Comprehension of Biological Complexity: Modeling, Execution, and Visualization of Thymic T-Cell Maturation. *Genome Research*, 13(11):2485–2497.

Egel, R. (1984a). The pedigree pattern of mating-type switching in Schizosaccharomycespombe. *Current Genetics*, 8(3):205–210.

Egel, R. (1984b). Two tightly linked silent cassettes in the mating-type region of Schizosaccharomyces pombe. *Current Genetics*, 8(3):199–203.

Elf, J. and Ehrenberg, M. (2004). Spontaneous separation of bi-stable biochemical systems into spatial domains of opposite phases. *Systems Biology, IEE Proceedings*, 1(2):230–236.

Ellner, S. P. and Guckenheimer, J. (2006). *Dynamic Models in Biology*. Princeton University Press.

Engquist, B., Lötstedt, P., and Runborg, O., editors (2009). *Multiscale Modeling and Simulation in Science*, volume 66 of *Lecture Notes in Computational Science and Engineering*. Springer.

Ewald, R. (2010). *Automatic Algorithm Selection for Complex Simulation Problems*. PhD thesis, University of Rostock.

Ewald, R., Himmelspach, J., Jeschke, M., Leye, S., and Uhrmacher, A. M. (2010). Flexible experimentation in the modeling and simulation framework JAMES II–implications for computational systems biology. *Briefings in Bioinformatics*, 11(3):290–300.

Ewald, R., Maus, C., Rolfs, A., and Uhrmacher, A. (2007). Discrete event modelling and simulation in systems biology. *Journal of Simulation*, 1(2):81–96.

Faeder, J. R. (2011). Toward a comprehensive language for biological systems. *BMC Biology*, 9:68.

Faeder, J. R., Blinov, M. L., Goldstein, B., and Hlavacek, W. S. (2005). Rule-Based Modeling of Biochemical Networks. *Complexity*, 10(4):22–41.

Faeder, J. R., Blinov, M. L., and Hlavacek, W. S. (2009). Rule-Based Modeling of Biochemical Systems with BioNetGen. In *Systems Biology*, volume 500 of *Methods in Molecular Biology*, pages 113–167. Humana Press.

Fages, F. and Soliman, S. (2008). Formal Cell Biology in Biocham. In Bernardo, M., Degano, P., and Zavattaro, G., editors, *Formal Methods for Computational Systems Biology*, volume 5016 of *Lecture Notes in Computer Science*, pages 54–80. Springer.

Feibleman, J. K. (1954). Theory of Integrative Levels. *British Journal for the Philosophy of Science*, 5(17):59–66.

Felleisen, M. (1991). On the expressive power of programming languages. *Science of Computer Programming*, 17(1–3):35–75.

Ferrell, J. E. (2009). Q&A: Systems biology. *Journal of Biology*, 8:2.

Fish, J., editor (2009). *Multiscale Methods: Bridging the Scales in Science and Engineering*. Oxford University Press.

Fisher, J. and Harel, D. (2010). On Statecharts for Biology. In Iyengar, M. S., editor, *Symbolic Systems Biology: Theory and Methods*, chapter 4. Jones & Bartlett Learning.

Fisher, J., Harel, D., and Henzinger, T. A. (2011). Biology as Reactivity. *Communications of the ACM*, 54(10):72–82.

Fishwick, P. A. and Lee, K. (1996). Two Methods For Exploiting Abstraction In Systems. In *AI, Simulation and Planning in High Autonomous Systems*, pages 257–264.

Fowler, M. (2004). *UML Distilled: A Brief Guide to the Standard Object Modeling Language*. Object Technology Series. Addison-Wesley Professional, 3rd edition.

Frantz, F. K. (1995). A taxonomy of model abstraction techniques. In Alexopoulos, C., Kang, K., Lilegdon, W. R., and Goldsman, D., editors, *Proceedings of the 1995 Winter Simulation Conference*, pages 1413–1420.

Gallagher, R. and Appenzeller, T. (1999). Beyond Reductionism. *Science*, 284(5411):79.

Gao, Q., Liu, F., Gilbert, D., Heiner, M., and Tree, D. (2011). A Multiscale Approach to Modelling Planar Cell Polarity in Drosophila Wing Using Hierarchically Coloured Petri Nets. In *Proceedings of the 9th International Conference on Computational Methods in Systems Biology*, pages 209–218. ACM.

Ghosh, R. and Tomlin, C. (2001). Lateral Inhibition through Delta-Notch Signaling: A Piecewise Affine Hybrid Model. In Di Benedetto, M. and Sangiovanni-Vincentelli, A., editors, *Hybrid Systems: Computation and Control*, volume 2034 of *Lecture Notes in Computer Science*, pages 232–246. Springer.

Giavitto, J.-L. and Michel, O. (2001). MGS: A Rule-Based Programming Language for Complex Objects and Collections. *Electronic Notes in Theoretical Computer Science*, 59(4):286–304.

Gilbert, D., Heiner, M., and Lehrack, S. (2007). A Unifying Framework for Modelling and Analysing Biochemical Pathways Using Petri Nets. In Calder, M. and Gilmore, S., editors, *Computational Methods in Systems Biology*, volume 4695 of *Lecture Notes in Computer Science*, pages 200–216. Springer.

Gille, C., Bölling, C., Hoppe, A., Bulik, S., Hoffmann, S., Hübner, K., Karl-städt, A., Ganeshan, R., König, M., Rother, K., Weidlich, M., Behre, J., and Holzhütter, H.-G. (2010). HepatoNet1: a comprehensive metabolic reconstruction of the human hepatocyte for the analysis of liver physiology. *Molecular Systems Biology*, 6:411.

Gillespie, D. T. (1977). Exact Stochastic Simulation of Coupled Chemical Reactions. *The Journal of Physical Chemistry*, 81(25):2340–2361.

Glanville, R. (1990). Sed Quis Custodient Ipsos Custodes? In Heylighen, F., Rosseel, E., and Demeyere, F., editors, *Self-Steering and Cognition in Complex Systems: Toward a New Cybernetics*, volume 22 of *Studies in Cybernetics*, pages 107–113. Gordon and Breach Science Publishers.

Goss, P. J. E. and Peccoud, J. (1998). Quantitative modeling of stochastic systems in molecular biology by using stochastic Petri nets. *Proceedings of the National Academy of Sciences*, 95(12):6750–6755.

Green, T. and Petre, M. (1992). When Visual Programs are Harder to Read than Textual Programs. In *Proceedings of the Sixth European Conference on Cognitive Ergonomics (ECCE-6), Budapest, Hungary*.

Green, T. and Petre, M. (1996). Usability Analysis of Visual Programming Environments: A 'Cognitive Dimensions' Framework. *Journal of Visual Languages and Computing*, 7:131–174.

Green, T., Petre, M., and Bellamy, R. (1991). Comprehensibility of Visual and Textual Programs: A Test of Superlativism Against the 'Match-Mismatch' Conjecture. In *Proceedings of Empirical Studies of Programmers: Fourth Workshop*, pages 121–146. Ablex Publishing Corporation.

Grene, M. (1969). Hierarchy: One Word, How Many Concepts? In Whyte, L., Wilson, A., and Wilson, D., editors, *Hierarchical structures*, pages 56–58. Elsevier, New York.

Gutfreund, H. (1995). *Kinetics for the Life Sciences: Receptors, Transmitters and Catalysts*. Cambridge University Press.

Hamm, H. E. (1998). The Many Faces of G Protein Signaling. *Journal of Biological Chemistry*, 273(2):669–672.

Harel, D. (1987). Statecharts: a visual formalism for complex systems. *Science of Computer Programming*, 8(3):231–274.

Harel, D. (1988). On visual formalisms. *Communications of the ACM*, 31(5):514–530.

Harel, D. and Gery, E. (1996). Executable Object Modeling with Statecharts. In *Proceedings of the 18th International Conference on Software Engineering*, pages 246–257. IEEE Computer Society.

Harel, D. and Kugler, H. (2004). The Rhapsody Semantics of Statecharts (or, On the Executable Core of the UML). In Ehrig, H., Damm, W., Desel, J., Große-Rhode, M., Reif, W., Schnieder, E., and Westkämper, E., editors, *Integration of Software Specification Techniques for Applications in Engineering*, volume 3147 of *Lecture Notes in Computer Science*, pages 325–354. Springer.

Harel, D. and Pnueli, A. (1985). On the Development of Reactive Systems. In Apt, K., editor, *Logics and Models of Concurrent Systems*, pages 477–498. Springer.

Harris, L., Hogg, J., and Faeder, J. (2009). Compartmental rule-based modeling of biochemical systems. In Rossetti, M., Hill, R., Johansson, B., Dunkin, A., and Ingalls, R., editors, *Proceedings of the 2009 Winter Simulation Conference*, pages 908–919.

Hartmanis, J. (1979). On the succinctness of different representations of languages. In Maurer, H., editor, *Automata, Languages and Programming*, volume 71 of *Lecture Notes in Computer Science*, pages 282–288. Springer.

Hayes-Roth, F. (1985). Rule-based systems. *Communications of the ACM*, 28(9):921–932.

Heiner, M. and Gilbert, D. (2011). How Might Petri Nets Enhance Your Systems Biology Toolkit. In Kristensen, L. and Petrucci, L., editors, *Appli-*

cations and Theory of Petri Nets, volume 6709 of *Lecture Notes in Computer Science*, pages 17–37. Springer.

Heiner, M., Gilbert, D., and Donaldson, R. (2008a). Petri Nets for Systems and Synthetic Biology. In Bernardo, M., Degano, P., and Zavattaro, G., editors, *Formal Methods for Computational Systems Biology*, volume 5016 of *Lecture Notes in Computer Science*, pages 215–264. Springer.

Heiner, M., Lehrack, S., Gilbert, D., and Marwan, W. (2009). Extended Stochastic Petri Nets for Model-Based Design of Wetlab Experiments. In Priami, C., Back, R.-J., and Petre, I., editors, *Transactions on Computational Systems Biology XI*, volume 5750 of *Lecture Notes in Computer Science*, pages 138–163. Springer.

Heiner, M., Richter, R., and Schwarick, M. (2008b). Snoopy – A Tool to Design and Animate/Simulate Graph-Based Formalisms. In *Proceedings of the International Workshop on Petri Nets Tools and APplications*.

Helms, T., Himmelspach, J., Maus, C., Röwer, O., Schützel, J., and Uhrmacher, A. (2012). Toward a language for the flexible observation of simulations. In Laroque, C., Himmelspach, J., Rasupathy, R., Rose, O., and Uhrmacher, A., editors, *Proceedings of the 2012 Winter Simulation Conference*, pages 3857–3868.

Henzinger, T., Jobstmann, B., and Wolf, V. (2009). Formalisms for Specifying Markovian Population Models. In Bournez, O. and Potapov, I., editors, *Reachability Problems*, volume 5797 of *Lecture Notes in Computer Science*, pages 3–23. Springer.

Heylighen, F. (1995). Downward Causation. In Heylighen, F., Joslyn, C., and Turchin, V., editors, *Principia Cybernetica Web*. Principia Cybernetica, Brussels. URL: http://cleamc11.vub.ac.be/DOWNCAUS.HTML, Accessed 10 November 2011.

Hillston, J. (1996). *A Compositional Approach to Performance Modelling*. Cambridge University Press.

Himmelspach, J., Ewald, R., and Uhrmacher, A. M. (2008). A flexible and scalable experimentation layer. In Mason, S. J., Hill, R. R., Mönch, L., Rose, O., Jefferson, T., and Fowler, J. W., editors, *Proceedings of the 2008 Winter Simulation Conference*, pages 827–835.

Himmelspach, J., Röhl, M., and Uhrmacher, A. M. (2010). Component-based models and simulations for supporting valid multi-agent system simulations. *Applied Artificial Intelligence*, 24(5):414–442.

Himmelspach, J. and Uhrmacher, A. (2007). Plug'n Simulate. In *Proceedings of the 40th Annual Simulation Symposium, ANSS '07*, pages 137–143.

Hirsch, S., Lloyd, B., Szczerba, D., and Székely, G. (2010). A Multiscale Simulation Framework for Modeling Solid Tumor Growth with an Explicit Vessel Network. In Deisboeck, T. S. and Stamatakos, G. S., editors, *Multiscale Cancer Modeling*, pages 309–337. CRC Press.

Hlavacek, W. S., Faeder, J. R., Blinov, M. L., Posner, R. G., Hucka, M., and Fontana, W. (2006). Rules for Modeling Signal-Transduction Systems. *Science STKE*, 2006(344):re6.

Hodgkin, A. L. and Huxley, A. F. (1952). A quantitative description of membrane current and its application to conduction and excitation in nerve. *Journal of Physiology*, 117(4):500–544.

Hoehme, S. and Drasdo, D. (2010). A cell-based simulation software for multicellular systems. *Bioinformatics*, 26(20):2641–2642.

Hogeweg, P. (2007). From population dynamics to ecoinformatics: Ecosystems as multilevel information processing systems. *Ecological Informatics*, 2(2):103–111.

Hogeweg, P. and Hesper, B. (1990). Individual-oriented modelling in ecology. *Mathematical and Computer Modelling*, 13(6):83–90.

Holcombe, M., Adra, S., Bicak, M., Chin, S., Coakley, S., Graham, A. I., Green, J., Greenough, C., Jackson, D., Kiran, M., MacNeil, S., Maleki-Dizaji, A., McMinn, P., Pogson, M., Poole, R., Qwarnstrom, E., Ratnieks,

F., Rolfe, M. D., Smallwood, R., Sun, T., and Worth, D. (2012). Modelling complex biological systems using an agent-based approach. *Integrative Biology*, 4(1):53–64.

Holte, R. C. and Choueiry, B. Y. (2003). Abstraction and reformulation in artificial intelligence. *Philosophical Transactions of the Royal Society of London, Series B: Biological Sciences*, 358(1435):1197–1204.

Hooper, J. W. (1986). Strategy-related characteristics of discrete-event languages and models. *SIMULATION*, 46(4):153–159.

Hoops, S., Sahle, S., Gauges, R., Lee, C., Pahle, J., Simus, N., Singhal, M., Xu, L., Mendes, P., and Kummer, U. (2006). COPASI–a COmplex PAthway SImulator. *Bioinformatics*, 22(24):3067–3074.

Huber, P., Jensen, K., and Shapiro, R. (1991). Hierarchies in Coloured Petri Nets. In Rozenberg, G., editor, *Advances in Petri Nets 1990*, volume 483 of *Lecture Notes in Computer Science*, pages 313–341. Springer.

Hucka, M., Finney, A., Sauro, H. M., Bolouri, H., Doyle, J. C., Kitano, H., Arkin, A. P., Bornstein, B. J., Bray, D., Cornish-Bowden, A., Cuellar, A. A., Dronov, S., Gilles, E. D., Ginkel, M., Gor, V., Goryanin, I. I., Hedley, W. J., Hodgman, T. C., Hofmeyr, J.-H., Hunter, P. J., Juty, N. S., Kasberger, J. L., Kremling, A., Kummer, U., Novère, N. L., Loew, L. M., Lucio, D., Mendes, P., Minch, E., Mjolsness, E. D., Nakayama, Y., Nelson, M. R., Nielsen, P. F., Sakurada, T., Schaff, J. C., Shapiro, B. E., Shimizu, T. S., Spence, H. D., Stelling, J., Takahashi, K., Tomita, M., Wagner, J., Wang, J., and Forum, S. B. M. L. (2003). The systems biology markup language (SBML): a medium for representation and exchange of biochemical network models. *Bioinformatics*, 19(4):524–531.

Hume, D. (1739). *Treatise of Human Nature, Book 1: Of the understanding.*

Hunter, P. J., Crampin, E. J., and Nielsen, P. M. F. (2008). Bioinformatics, multiscale modeling and the IUPS Physiome Project. *Briefings in Bioinformatics*, 9(4):333–343.

Hunter, P. J., Pullan, A. J., and Smaill, B. H. (2003). Modeling Total Heart Function. *Annual Review of Biomedical Engineering*, 5:147–177.

Ideker, T., Galitski, T., and Hood, L. (2001). A New Approach to Decoding Life: Systems Biology. *Annual Review of Genomics and Human Genetics*, 2:343–372.

Imai, Y. and Yamamoto, M. (1992). Schizosaccharomyces pombe sxa1+ and sxa2+ encode putative proteases involved in the mating response. *Molecular and Cellular Biology*, 12(4):1827–1834.

Iordache, O. (2011). *Modeling Multi-Level Systems*. Springer.

Jackson, R. A. (2004). *Mechanisms in Organic Reactions*, volume 23 of *Tutorial Chemistry Texts*. Royal Society of Chemistry.

Janowski, S., Kormeier, B., Töpel, T., Hippe, K., Hofestädt, R., Willassen, N., Friesen, R., Rubert, S., Borck, D., Haugen, P., and Chen, M. (2010). Modeling of Cell-to-Cell Communication Processes with Petri Nets Using the Example of Quorum Sensing. *In Silico Biology*, 10(1–2):27–48.

Jensen, K. (1998). An introduction to the practical use of coloured Petri Nets. In Reisig, W. and Rozenberg, G., editors, *Lectures on Petri Nets II: Applications*, volume 1492 of *Lecture Notes in Computer Science*, pages 237–292. Springer.

Jensen, K. and Kristensen, L. M. (2009). *Coloured Petri Nets: Modeling and Validation of Concurrent Systems*. Springer.

Jeschke, M., Ewald, R., and Uhrmacher, A. M. (2011). Exploring the performance of spatial stochastic simulation algorithms. *Journal of Computational Physics*, 230(7):2562–2574.

John, M. (2010). *Reaction Constraints for the Pi-Calculus: A Language for the Stochastic and Spatial Modeling of Cell-Biological Processes*. PhD thesis, University of Rostock.

John, M., Lhoussaine, C., and Niehren, J. (2009). Dynamic Compartments in the Imperative π-Calculus. In Degano, P. and Gorrieri, R., editors, *Computational Methods in Systems Biology*, volume 5688 of *Lecture Notes in Computer Science*, pages 235–250. Springer.

John, M., Lhoussaine, C., Niehren, J., and Uhrmacher, A. (2008). The Attributed Pi Calculus. In Heiner, M. and Uhrmacher, A., editors, *Computational Methods in Systems Biology*, volume 5307 of *Lecture Notes in Computer Science*, pages 83–102. Springer.

John, M., Lhoussaine, C., Niehren, J., and Uhrmacher, A. (2010). The Attributed Pi-Calculus with Priorities. In Priami, C., Breitling, R., Gilbert, D., Heiner, M., and Uhrmacher, A., editors, *Transactions on Computational Systems Biology XII*, volume 5945 of *Lecture Notes in Computer Science*, pages 13–76. Springer.

John, M., Lhoussaine, C., Niehren, J., and Versari, C. (2011). Biochemical Reaction Rules with Constraints. In Barthe, G., editor, *Programming Languages and Systems: 20th European Symposium on Programming, ESOP 2011*, volume 6602 of *Lecture Notes in Computer Science*, pages 338–357. Springer.

Johnston, L. A. and Edgar, B. A. (1998). Wingless and Notch regulate cell-cycle arrest in the developing Drosophila wing. *Nature*, 394(6688):82–84.

Kam, N., Cohen, I., and Harel, D. (2001). The Immune System as a Reactive System: Modeling T Cell Activation With Statecharts. In *Proceedings of the IEEE Symposia on Human-Centric Computing Languages and Environments*, pages 15–22.

Kar, S., Baumann, W. T., Paul, M. R., and Tyson, J. J. (2009). Exploring the roles of noise in the eukaryotic cell cycle. *Proceedings of the National Academy of Sciences of the United States of America*, 106(16):6471–6476.

Kauffman, S. A. (2008). *Reinventing the Sacred: A New View of Science, Reason, and Religion*. Basic Books.

Keane, J. (2003). Tools for modelling biological processes. *Nature*, 421(6923):573.

Kesten, Y. and Pnueli, A. (1991). Timed and Hybrid Statecharts and their textual representation. In Vytopil, J., editor, *Formal Techniques in Real-Time and Fault-Tolerant Systems*, volume 571 of *Lecture Notes in Computer Science*, pages 591–620. Springer.

Kherlopian, A. R., Song, T., Duan, Q., Neimark, M. A., Po, M. J., Gohagan, J. K., and Laine, A. F. (2008). A review of imaging techniques for systems biology. *BMC Systems Biology*, 2:74.

Kim, S. M. and Huberman, J. A. (2001). Regulation of replication timing in fission yeast. *The EMBO Journal*, 20(21):6115–6126.

Kirschner, M. W. (2005). The Meaning of Systems Biology. *Cell*, 121(4):503–504.

Kitano, H. (2002a). Computational systems biology. *Nature*, 420(6912):206–210.

Kitano, H. (2002b). Systems Biology: A Brief Overview. *Science*, 295(5560):1662–1664.

Klar, A. J. (1992). Developmental choices in mating-type interconversion in fission yeast. *Trends in Genetics*, 8(6):208–213.

Klar, A. J., Bonaduce, M. J., and Cafferkey, R. (1991). The Mechanism of Fission Yeast Mating Type Interconversion: Seal/Replicate/Cleave Model of Replication Across the Double-Stranded Break Site at mat1. *Genetics*, 127(3):489–496.

Klipp, E., Herwig, R., Kowald, A., Wierling, C., and Lehrach, H. (2005). *Systems Biology in Practice: Concepts, Implementation and Application.* Wiley-VCH, Weinheim.

Kohl, P., Crampin, E., Quinn, T., and Noble, D. (2010). Systems Biology: An Approach. *Clinical Pharmacology & Therapeutics*, 88(1):25–33.

Korn, G. A. and Wait, J. V. (1978). *Digital Continuous-System Simulation.* Prentice-Hall.

Kreft, I. and de Leeuw, J. (1998). *Introducing Multilevel Modeling.* Sage Publications.

Krivine, J., Milner, R., and Troina, A. (2008). Stochastic Bigraphs. *Electronic Notes in Theoretical Computer Science*, 218:73–96. Proceedings of the 24th Conference on the Mathematical Foundations of Programming Semantics (MFPS XXIV).

Kugler, H., Larjo, A., and Harel, D. (2010). Biocharts: a visual formalism for complex biological systems. *Journal of The Royal Society Interface*, 7(48):1015–1024.

Kummer, O. (2001). Introduction to Petri nets and reference nets. *Sozionik Aktuell*, 1:1–9.

Kuttler, C., Lhoussaine, C., and Nebut, M. (2010). Rule-Based Modeling of Transcriptional Attenuation at the Tryptophan Operon. In Priami, C., Breitling, R., Gilbert, D., Heiner, M., and Uhrmacher, A., editors, *Transactions on Computational Systems Biology XII*, volume 5945 of *Lecture Notes in Computer Science*, pages 199–228. Springer.

Kuttler, C. and Niehren, J. (2006). Gene Regulation in the Pi Calculus: Simulating Cooperativity at the Lambda Switch. In Priami, C., Ingólfsdóttir, A., Mishra, B., and Riis Nielson, H., editors, *Transactions on Computational Systems Biology VII*, volume 4230 of *Lecture Notes in Computer Science*, pages 24–55. Springer.

Köhler-Bußmeier, M. (2009). Hornets: Nets within Nets Combined with Net Algebra. In Franceschinis, G. and Wolf, K., editors, *Applications and Theory of Petri Nets*, volume 5606 of *Lecture Notes in Computer Science*, pages 243–262. Springer.

Lai, E. C. (2004). Notch signaling: control of cell communication and cell fate. *Development*, 131(5):965–973.

Lai, X., Nikolov, S., Wolkenhauer, O., and Vera, J. (2009). A multi-level model accounting for the effects of JAK2-STAT5 signal modulation in erythropoiesis. *Comput Biol Chem*, 33(4):312–324.

Lakoff, G. (1987). *Women, Fire, and Dangerous Things: What Categories Reveal About the Mind.* University Of Chicago Press.

Laneve, C. and Tarissan, F. (2008). A simple calculus for proteins and cells. *Theoretical Computer Science*, 404(1–2):127–141.

Lavelle, C., Berry, H., Beslon, G., Ginelli, F., Giavitto, J.-L., Kapoula, Z., Le Bivic, A., Peyrieras, N., Radulescu, O., Six, A., Thomas-Vaslin, V., and Bourgine, P. (2008). From Molecules to Organisms: Towards Multiscale Integrated Models of Biological Systems. *Theoretical Biology Insights*, 1:13–22.

Le Novère, N., Bornstein, B., Broicher, A., Courtot, M., Donizelli, M., Dharuri, H., Li, L., Sauro, H., Schilstra, M., Shapiro, B., Snoep, J. L., and Hucka, M. (2006). BioModels Database: a free, centralized database of curated, published, quantitative kinetic models of biochemical and cellular systems. *Nucleic Acids Research*, 34(Database issue):D689–D691.

Le Novère, N., Hucka, M., Mi, H., Moodie, S., Schreiber, F., Sorokin, A., Demir, E., Wegner, K., Aladjem, M. I., Wimalaratne, S. M., Bergman, F. T., Gauges, R., Ghazal, P., Kawaji, H., Li, L., Matsuoka, Y., Villéger, A., Boyd, S. E., Calzone, L., Courtot, M., Dogrusoz, U., Freeman, T. C., Funahashi, A., Ghosh, S., Jouraku, A., Kim, S., Kolpakov, F., Luna, A., Sahle, S., Schmidt, E., Watterson, S., Wu, G., Goryanin, I., Kell, D. B., Sander, C., Sauro, H., Snoep, J. L., Kohn, K., and Kitano, H. (2009). The Systems Biology Graphical Notation. *Nature Biotechnology*, 27(8):735–741.

Lee, J., Niederer, S., Nordsletten, D., Grice, I. L., Smaill, B., Kay, D., and Smith, N. (2009). Coupling contraction, excitation, ventricular and coronary blood flow across scale and physics in the heart. *Philosophical Transactions of the Royal Society A*, 367(1896):2311–2331.

Lee, K. and Fishwick, P. A. (1996). Dynamic model abstraction. In Charnes, J., Morrice, D., Brunner, D., and Swain, J., editors, *Proceedings of the 1996 Winter Simulation Conference*, pages 764–771.

Leitch, R., Shen, Q., Coghill, G., and Chantler, M. (1999). Choosing the right model. *IEE Proceedings – Control Theory & Applications*, 146(5):435–449.

Leupold, U. (1958). Studies on recombination in Schizosaccharomyces pombe. *Cold Spring Harbor Symposia on Quantitative Biology*, 23:161–170.

Lewis, J. (2003). Autoinhibition with Transcriptional Delay. *Current Biology*, 13(16):1398–1408.

Li, C., Donizelli, M., Rodriguez, N., Dharuri, H., Endler, L., Chelliah, V., Li, L., He, E., Henry, A., Stefan, M. I., Snoep, J. L., Hucka, M., Le Novère, N., and Laibe, C. (2010). BioModels Database: An enhanced, curated and annotated resource for published quantitative kinetic models. *BMC Systems Biology*, 4:92.

Lomazova, I. A. (2000). Nested Petri Nets – a Formalism for Specification and Verification of Multi-Agent Distributed Systems. *Fundamenta Informaticae*, 43(1–4):195–214.

Lomazova, I. A. (2001). Nested Petri Nets: Multi-level and Recursive Systems. *Fundamenta Informaticae*, 47(3–4):283–293.

Luke, S., Cioffi-Revilla, C., Panait, L., Sullivan, K., and Balan, G. (2005). MASON: A Multiagent Simulation Environment. *SIMULATION*, 81(7):517–527.

Luna, J. J. (1993). Hierarchical Relations in Simulation Models. In Evans, G., Mollaghasemi, M., Russell, E., and Biles, W., editors, *Proceedings of the 1993 Winter Simulation Conference*, pages 132–137.

Lynch, S. (2004). *Dynamical systems with applications using MATLAB*. Birkhäuser.

Machado, D., Costa, R., Rocha, M., Ferreira, E., Tidor, B., and Rocha, I. (2011). Modeling formalisms in systems biology. *AMB Express*, 1:45.

MacNamara, S. and Burrage, K. (2010). Stochastic Modeling of Naïve T Cell Homeostasis for Competing Clonotypes via the Master Equation. *Multiscale Modeling & Simulation*, 8(4):1325–1347.

MacNamara, S. and Burrage, K. (2011). Stochastic Modelling of T Cell Homeostasis for Two Competing Clonotypes Via the Master Equation. In Molina-París, C. and Lythe, G., editors, *Mathematical Models and Immune Cell Biology*, pages 207–225. Springer.

MacNamara, S., Burrage, K., and Sidje, R. B. (2008). Multiscale Modeling of Chemical Kinetics via the Master Equation. *Multiscale Modeling & Simulation*, 6(4):1146–1168.

Marsan, M. A., Balbo, G., Conte, G., Donatelli, S., and Franceschinis, G. (1995). *Modelling With Generalized Stochastic Petri Nets*. Wiley Series in Parallel Computing. John Wiley & Sons.

Marwan, W., Wagler, A., and Weismantel, R. (2009). Petri nets as a framework for the reconstruction and analysis of signal transduction pathways and regulatory networks. *Natural Computing*, 10(2):639–654.

Matsuno, H., Doi, A., Nagasaki, M., and Miyano, S. (2000). Hybrid Petri net representation of gene regulatory network. *Pacific Symposium on Biocomputing*, 5:338–349.

Matsuno, H., Tanaka, Y., Aoshima, H., Doi, A., Matsui, M., and Miyano, S. (2003). Biopathways Representation and Simulation on Hybrid Functional Petri Net. *In Silico Biology*, 3(3):389–404.

Maus, C. (2008). Component-Based Modelling of RNA Structure Folding. In Heiner, M. and Uhrmacher, A., editors, *Computational Methods in Systems Biology*, volume 5307 of *Lecture Notes in Computer Science*, pages 44–62. Springer.

Maus, C., John, M., Röhl, M., and Uhrmacher, A. M. (2008). Hierarchical Modeling for Computational Biology. In Bernardo, M., Degano, P., and Zavattaro, G., editors, *Formal Methods for Computational Systems Biology*, volume 5016 of *Lecture Notes in Computer Science*, pages 81–124. Springer.

Maus, C., Rybacki, S., and Uhrmacher, A. M. (2011). Rule-based multi-level modeling of cell biological systems. *BMC Systems Biology*, 5:166.

Mazemondet, O., John, M., Maus, C., Uhrmacher, A., and Rolfs, A. (2009). Integrating diverse reaction types into stochastic models – A signaling pathway case study in the Imperative π-Calculus. In Rossetti, M., Hill, R., Johansson, B., Dunkin, A., and Ingalls, R., editors, *Proceedings of the 2009 Winter Simulation Conference*, pages 932–943.

McCarthy, J. and Hayes, P. J. (1969). Some Philosophical Problems from the Standpoint of Artificial Intelligence. In Meltzer, B. and Michie, D., editors, *Machine Intelligence 4*, pages 463–502. Edinburgh University Press.

Meier-Schellersheim, M., Fraser, I. D. C., and Klauschen, F. (2009). Multiscale modeling for biologists. *Wiley Interdisciplinary Reviews: Systems Biology and Medicine*, 1(1):4–14.

Meineke, F., Potten, C., and Loeffler, M. (2001). Cell migration and organization in the intestinal crypt using a lattice-free model. *Cell Proliferation*, 34(4):253–266.

Meinhardt, H. and Gierer, A. (2000). Pattern formation by local self-activation and lateral inhibition. *BioEssays*, 22:753–760.

Mesarović, M. D. (1968). Systems Theory and Biology–View of a Theoretician. In Mesarović, M. D., editor, *Systems Theory and Biology: Proceedings of the III Systems Symposium at Case Institute of Technology*, pages 59–87. Springer, Berlin, Heidelberg, New York.

Mesarović, M. D., Macko, D., and Takahara, Y. (1970). *Theory of Hierarchical, Multilevel Systems*. Academic Press.

Michel, O., Spicher, A., and Giavitto, J.-L. (2009). Rule-based programming for integrative biological modeling. *Natural Computing*, 8(4):865–889.

Miller, J. G. (1978). *Living systems*. McGraw-Hill.

Milner, R. (1980). *A Calculus of Communicating Systems*. Springer.

Milner, R. (1999). *Communicating and Mobile Systems: the π-Calculus*. Cambridge University Press.

Milner, R. (2001). Bigraphical Reactive Systems. In Larsen, K. and Nielsen, M., editors, *CONCUR 2001 – Concurrency Theory*, volume 2154 of *Lecture Notes in Computer Science*, pages 16–35. Springer.

Milner, R. (2002). Bigraphs as a Model for Mobile Interaction. In Corradini, A., Ehrig, H., Kreowski, H., and Rozenberg, G., editors, *Graph Transformation*, volume 2505 of *Lecture Notes in Computer Science*, pages 8–13. Springer.

Milner, R. (2006). Pure bigraphs: Structure and dynamics. *Information and Computation*, 204(1):60–122.

Milner, R. (2009). *The Space and Motion of Communicating Agents*. Cambridge University Press.

Milner, R., Parrow, J., and Walker, D. (1992). A Calculus of Mobile Processes, I. *Information and Computation*, 100(1):1–40.

Minsky, M. (1965). Matter, Mind and Models. In Kalenich, W. A., editor, *Proceedings of IFIP Congress 65. Organized by the International Federation for Information Processing*, pages 45–49. Spartan Books, Washington D.C.

Miyamoto, T. and Kumagai, S. (2005). A Survey of Object-Oriented Petri Nets and Analysis Methods. *IEICE Transactions on Fundamentals of Electronics, Communications and Computer Sciences*, E88-A(11):2964–2971.

Miyata, H., Miyata, M., and Ito, M. (1978). The cell cycle in the fission yeast, Schizosaccharomyces pombe. I. Relationship between cell size and cycle time. *Cell Structure and Function*, 3(1):39–46.

M'Mahon, J. H. (1857). *The Metaphysics of Aristotle*. Henry G. Bohn, London.

Moher, T., Mak, D., Blumenthal, B., and Levanthal, L. (1993). Comparing the Comprehensibility of Textual and Graphical Programs: The Case of Petri Nets. In *Proceedings of Empirical Studies of Programmers: Fifth Workshop*, pages 137–161. Ablex Publishing Corporation.

Moseley, J. B., Mayeux, A., Paoletti, A., and Nurse, P. (2009). A spatial gradient coordinates cell size and mitotic entry in fission yeast. *Nature*, 459(7248):857–860.

Mura, I., Prandi, D., Priami, C., and Romanel, A. (2009). Exploiting non-Markovian Bio-Processes. *Electronic Notes in Theoretical Computer Science*, 253(3):83–98. Proceedings of Seventh Workshop on Quantitative Aspects of Programming Languages (QAPL 2009).

Möhring, M. (1996). Social Science Multilevel Simulation with MIMOSE. In Troitzsch, K., Mueller, U., Gilbert, G., and Doran, J., editors, *Social Science Microsimulation*, pages 123–137. Springer, Berlin, Heidelberg, New York.

Neves, S. R. and Iyengar, R. (2009). Models of Spatially Restricted Biochemical Reaction Systems. *Journal of Biological Chemistry*, 284(9):5445–5449.

Nielsen, O. and Davey, J. (1995). Pheromone communication in the fission yeast Schizosaccharomyces pombe. *Seminars in Cell Biology*, 6(2):95–104.

Noble, D. (2002). Modeling the Heart–from Genes to Cells to the Whole Organ. *Science*, 295(5560):1678–1682.

Noble, D. (2006). *The Music of Life: Biology Beyond the Genome.* Oxford University Press.

Noble, D. (2008). Claude Bernard, the first systems biologist, and the future of physiology. *Experimental Physiology*, 93(1):16–26.

Novak, B. and Tyson, J. J. (1997). Modeling the control of DNA replication in fission yeast. *Proceedings of the National Academy of Sciences of the United States of America*, 94(17):9147–9152.

Oury, N. and Plotkin, G. D. (2011). Coloured Stochastic Multilevel Multiset Rewriting. In *Proceedings of the 9th International Conference on Computational Methods in Systems Biology*, pages 171–181.

Owen, M. R. and Sherratt, J. A. (1998). Mathematicalmodelling of juxtacrine-cellsignalling. *Mathematical Biosciences*, 153(2):125–150.

Paige, R., Ostroff, J., and Brooke, P. (2000). Principles for modeling language design. *Information and Software Technology*, 42(10):665–675.

Pedersen, M. and Plotkin, G. (2010). A Language for Biochemical Systems: Design and Formal Specification. In Priami, C., Breitling, R., Gilbert, D., Heiner, M., and Uhrmacher, A., editors, *Transactions on Computational Systems Biology XII*, volume 5945 of *Lecture Notes in Computer Science*, pages 77–145. Springer.

Petre, M. (1995). Why Looking Isn't Always Seeing: Readership Skills and Graphical Programming. *Communications of the ACM*, 38(6):33–44.

Petri, C. A. and Reisig, W. (2008). Petri net. *Scholarpedia*, 3(4):6477. URL: http://dx.doi.org/10.4249/scholarpedia.6477, Accessed 24 January 2012.

Phillips, A. (2009). An Abstract Machine for the Stochastic Bioambient calculus. *Electronic Notes in Theoretical Computer Science*, 227:143–159.

Phillips, A. and Cardelli, L. (2004). A Correct Abstract Machine for the Stochastic Pi-calculus. In *Concurrent Models in Molecular Biology, Bio-CONCUR*.

Phillips, A., Cardelli, L., and Castagna, G. (2006). A Graphical Representation for Biological Processes in the Stochastic pi-Calculus. In Priami, C., Ingólfsdóttir, A., Mishra, B., and Riis Nielson, H., editors, *Transactions on Computational Systems Biology VII*, volume 4230 of *Lecture Notes in Computer Science*, pages 123–152. Springer.

Polanyi, M. (1968). Life's Irreducible Structure. *Science*, 160(3834):1308–1312.

Pourquié, O. (2003). The Segmentation Clock: Converting Embryonic Time into Spatial Pattern. *Science*, 301(5631):328–330.

Pradal, C., Dufour-Kowalski, S., Boudon, F., Fournier, C., and Godin, C. (2008). OpenAlea: a visual programming and component-based software platform for plant modelling. *Functional Plant Biology*, 35(4086):751–760.

Priami, C. (1995). Stochastic π-Calculus. *The Computer Journal*, 38(7):578–589.

Priami, C., Regev, A., Shapiro, E., and Silverman, W. (2001). Application of a stochastic name-passing calculus to representation and simulation of molecular processes. *Information Processing Letters*, 80(1):25–31.

Proß, S. and Bachmann, B. (2011). An Advanced Environment for Hybrid Modeling of Biological Systems Based on Modelica. *Journal of Integrative Bioinformatics*, 8(1):152.

Pumain, D. (2006). *Hierarchy in Natural and Social Sciences*. Springer Science & Business.

Păun, G. and Rozenberg, G. (2002). A guide to membrane computing. *Theoretical Computer Science*, 287(1):73–100.

Qu, Z., MacLellan, W. R., and Weiss, J. N. (2003). Dynamics of the Cell Cycle: Checkpoints, Sizers, and Timers. *Biophysical Journal*, 85(6):3600–3611.

Reddy, V. N., Mavrovouniotis, M. L., and Liebman, M. N. (1993). Petri Net Representations in Metabolic Pathways. In *Proceedings of the International Conference on Intelligent Systems for Molecular Biology*, volume 1, pages 328–336.

Regev, A. (2002). *Computational Systems Biology: A Calculus for Biomolecular knowledge*. PhD thesis, Tel Aviv University.

Regev, A., Panina, E. M., Silverman, W., Cardelli, L., and Shapiro, E. (2004). BioAmbients: an abstraction for biological compartments. *Theoretical Computer Science*, 325(1):141–167.

Regev, A., Silverman, W., and Shapiro, E. (2001). Representation and simulation of biochemical processes using the π-calculus process algebra. In *Pacific Symposium on Biocomputing*, volume 6, pages 459–470.

Ribba, B., Colin, T., and Schnell, S. (2006). A multiscale mathematical model of cancer, and its use in analyzing irradiation therapies. *Theoretical Biology and Medical Modelling*, 3:7.

Risco-Martín, J., Mittal, S., Zeigler, B., and de la Cruz, J. (2007). From UML State Charts to DEVS State Machines using XML. In *Proceedings Workshop on Multi-Paradigm Modeling at the ACM/IEEE 10th International Conference on Model-Driven Engineering Languages and Systems*, pages 35–48.

Rohr, C., Marwan, W., and Heiner, M. (2010). Snoopy–a unifying Petri net framework to investigate biomolecular networks. *Bioinformatics*, 26(7):974–975.

Röhl, M. (2008). *Definition und Realisierung einer Plattform zur modellbasierten Komposition von Simulationsmodellen*. PhD thesis, University of Rostock, Rostock, Germany.

Salthe, S. N. (1985). *Evolving Hierarchical Systems: Their Structure and Representation*. Columbia University Press.

Salthe, S. N. (1989). Self-organization of/in Hierarchically Structured Systems. *Systems Research*, 6(3):199–208.

Salthe, S. N. (1991). Two forms of hierarchy theory in western discourses. *International Journal of General Systems*, 18(3):251–264.

Salthe, S. N. (1993). *Development and Evolution: Complexity and Change in Biology*. MIT Press.

Salthe, S. N. (2001). Summary of the Principles of Hierarchy Theory. On-line paper. URL: http://www.nbi.dk/~natphil/salthe/Summary_of_the_Principles_o.pdf, Accessed 08 November 2011.

Salthe, S. N. and Matsuno, K. (1995). Self-Organization in Hierarchical Systems. *Journal of Social and Evolutionary Systems*, 18(4):327–338.

Sauro, H. M. (2011). *Enzyme Kinetics for Systems Biology*. Future Skill Software.

Sauro, H. M., Uhrmacher, A. M., Harel, D., Hucka, M., Kwiatkowska, M., Mendes, P., Shaffer, C. A., Strömback, L., and Tyson, J. J. (2006). Challenges for Modeling and Simulation Methods in Systems Biology. In Perrone,

L., Wieland, F., Liu, J., Lawson, B., Nicol, D., and Fujimoto, R., editors, *Proceedings of the 2006 Winter Simulation Conference*, pages 1720–1730.

Sawin, K. E. (2009). Cell cycle: Cell division brought down to size. *Nature*, 459(7248):782–783.

Schillo, M., Fischer, K., and Klein, C. (2001). The Micro-Macro Link in DAI and Sociology. In Moss, S. and Davidsson, P., editors, *Multi-Agent-Based Simulation*, volume 1979/2001 of *Lecture Notes in Computer Science*, pages 303–317. Springer.

Schlegel, R. (1961). Mario Bunge on Causality. *Philosophy of Science*, 28(1):72–82.

Schweisguth, F. (2004). Regulation of Notch Signaling Activity. *Current Biology*, 14(3):R129–R138.

Shampine, L. F. and Thompson, S. (2007). Stiff systems. *Scholarpedia*, 2(3):2855. URL: http://dx.doi.org/10.4249/scholarpedia.2855, Accessed 16 February 2012.

Simon, H. A. (1962). The Architecture of Complexity. *Proceedings of the American Philosophical Society*, 106(6):467–482.

Simon, H. A. (1973). The Organization of Complex Systems. In Pattee, H., editor, *Hierarchy Theory – The Challenge of Complex Systems*, pages 1–27. George Braziller, New York.

Smith, J. M. and Smith, D. C. P. (1977). Database Abstractions: Aggregation and Generalization. *ACM Transactions on Database Systems*, 2(2):105–133.

Sneddon, M., Faeder, J., and Emonet, T. (2011). Efficient modeling, simulation and coarse-graining of biological complexity with NFsim. *Nature Methods*, 8(2):177–183.

Snijders, T. A. B. and Bosker, R. J. (1999). *Multilevel Analysis: An introduction to basic and advanced multilevel modeling*. Sage Publications.

Sorger, P. K. (2005). A reductionist's systems biology: Opinion. *Current Opinion in Cell Biology*, 17(1):9–11.

Spicher, A., Michel, O., Cieslak, M., Giavitto, J.-L., and Prusinkiewicz, P. (2008). Stochastic P systems and the simulation of biochemical processes with dynamic compartments. *BioSystems*, 91(3):458–472.

Sriuranpong, V., Borges, M. W., Ravi, R. K., Arnold, D. R., Nelkin, B. D., Baylin, S. B., and Ball, D. W. (2001). Notch Signaling Induces Cell Cycle Arrest in Small Cell Lung Cancer Cells. *Cancer Research*, 61(7):3200–3205.

Stamatakos, G. S. (2010). *In Silico* Oncology: Part I–Clinically Oriented Cancer Multilevel Modeling Based on Discrete Event Simulation. In Deisboeck, T. S. and Stamatakos, G. S., editors, *Multiscale Cancer Modeling*, pages 407–436. CRC Press.

Starfield, A. M. (1990). Qualitative, rule-based modeling. *BioScience*, 40(8):601–604.

Steiniger, A., Krüger, F., and Uhrmacher, A. M. (2012). Modeling Agents and their Environment in Multi-Level-DEVS. In Laroque, C., Himmelspach, J., Rasupathy, R., Rose, O., and Uhrmacher, A., editors, *Proceedings of the 2012 Winter Simulation Conference*, pages 2629–2640.

Stern, B. and Nurse, P. (1997). Fission yeast pheromone blocks S-phase by inhibiting the G1 cyclin B-p34cdc2 kinase. *The EMBO Journal*, 16(3):534–544.

Sun, T., Adra, S., Smallwood, R., Holcombe, M., and MacNeil, S. (2009). Exploring hypotheses of the actions of TGF-β1 in epidermal wound healing using a 3D computational multiscale model of the human epidermis. *PLoS One*, 4(12):e8515.

Sveiczer, A., Tyson, J. J., and Novak, B. (2001). A stochastic, molecular model of the fission yeast cell cycle: role of the nucleocytoplasmic ratio in cycle time regulation. *Biophysical Chemistry*, 92(1-2):1–15.

Takahara, Y. and Shiba, N. (1996). Systems theory and systems implementation-case of dss. In Klir, G. and Ören, T., editors, *Computer Aided Systems Theory – CAST '94*, volume 1105 of *Lecture Notes in Computer Science*, pages 388–408. Springer.

Takahashi, K., Kaizu, K., Hu, B., and Tomita, M. (2004). A multi-algorithm, multi-timescale method for cell simulation. *Bioinformatics*, 20(4):538–546.

Tilly, C. (1998). Micro, Macro, or Megrim? In Schlumbohm, J., editor, *Mikrogeschichte, Makrogeschichte: komplementär oder inkommensurabel?*, volume 7 of *Göttinger Gespräche zur Geschichtswissenschaft*, pages 33–51. Wallstein Verlag.

Timpf, S. (1999). Abstraction, Levels of Detail, and Hierarchies in Map Series. In Freksa, C. and Mark, D., editors, *Spatial Information Theory. Cognitive and Computational Foundations of Geographic Information Science*, volume 1661 of *Lecture Notes in Computer Science*, pages 125–140. Springer.

Tisue, S. and Wilensky, U. (2004). NetLogo: A Simple Environment for Modeling Complexity. In *Proceedings of the International Conference on Complex Systems*, pages 16–21.

Troitzsch, K. G., Mueller, U., Gilbert, G. N., and Doran, J. E., editors (1996). *Social Science Microsimulation*. Springer, Berlin, Heidelberg, New York.

Twycross, J., Band, L., Bennett, M., King, J., and Krasnogor, N. (2010). Stochastic and deterministic multiscale models for systems biology: an auxin-transport case study. *BMC Systems Biology*, 4:34.

Tyson, J. J. (1991). Modeling the cell division cycle: cdc2 and cyclin interactions. *Proceedings of the National Academy of Sciences of the United States of America*, 88(16):7328–7332.

Tyson, J. J., Csikasz-Nagy, A., and Novak, B. (2002). The dynamics of cell cycle regulation. *Bioessays*, 24(12):1095–1109.

Tyson, J. J. and Novak, B. (2001). Regulation of the Eukaryotic Cell Cycle: Molecular Antagonism, Hysteresis, and Irreversible Transitions. *Journal of Theoretical Biology*, 210(2):249–263.

Uhrmacher, A. (1992). *EMSY – Ein Modellierungskonzept für ökologische und biologische Systeme unter besonderer Berücksichtigung ihrer dynamischen Veränderung*, volume 20 of *Dissertationen zur künstlichen Intelligenz*. Infix.

Uhrmacher, A., Himmelspach, J., Röhl, M., and Ewald, R. (2006). Introducing Variable Ports and Multi-Couplings for Cell Biological Modeling in DEVS. In Perrone, L., Wieland, F., Liu, J., Lawson, B., Nicol, D., and Fujimoto, R., editors, *Proceedings of the 2006 Winter Simulation Conference*, pages 832–840.

Uhrmacher, A. and Kuttler, C. (2006). Multi-Level Modeling in Systems Biology by Discrete Event Approaches. *it – Information Technology*, 48(3):148–153.

Uhrmacher, A. M. (1993). Variable Structure Models: Autonomy and Control - Answers from Two Different Modeling Approaches. In *AI, Simulation, and Planning in High Autonomy Systems, 1993. Integrating Virtual Reality and Model-Based Environments. Proceedings. Fourth Annual Conference*, pages 133–139.

Uhrmacher, A. M. (1995). Reasoning About Changing Structure: A Modeling Concept for Ecological Systems. *Applied Artificial Intelligence*, 9(2):157–180.

Uhrmacher, A. M. (2001). Dynamic Structures in Modeling and Simulation: A Reflective Approach. *ACM Transactions on Modeling and Computer Simulation (TOMACS)*, 11(2):206–232.

Uhrmacher, A. M., Degenring, D., and Zeigler, B. (2005). Discrete Event Multi-level Models for Systems Biology. In Priami, C., editor, *Transactions on Computational Systems Biology*, volume 3380 of *Lecture Notes in Computer Science*, pages 133–133. Springer.

Uhrmacher, A. M., Ewald, R., John, M., Maus, C., Jeschke, M., and Biermann, S. (2007). Combining micro and macro-modeling in DEVS for computational biology. In Henderson, S., Biller, B., Hsieh, M.-H., Shortle, J., Tew, J., and

Barton, R., editors, *Proceedings of the 2007 Winter Simulation Conference*, pages 871–880. IEEE Press.

Uhrmacher, A. M. and Priami, C. (2005). Discrete event systems specification in systems biology – a discussion of stochastic pi calculus and DEVS. In Kuhl, M., Steiger, N., Armstrong, F., and Joines, J. A., editors, *Proceedings of the 2005 Winter Simulation Conference*, pages 317–326.

Uhrmacher, A. M. and Zeigler, B. P. (1996). Variable structure models in object-oriented simulation. *International Journal of General Systems*, 24(4):359–375.

Ullah, M. and Wolkenhauer, O. (2011). *Stochastic Approaches for Systems Biology*. Springer.

Valk, R. (2004). Object Petri Nets: Using the Nets-within-Nets Paradigm. In Desel, J., Reisig, W., and Rozenberg, G., editors, *Lectures on Concurrency and Petri Nets*, volume 3098 of *Lecture Notes in Computer Science*, pages 819–848. Springer.

van der Hoef, M., Ye, M., van Sint Annaland, M., Andrews, A., Sundaresan, S., and Kuipers, J. (2006). Multiscale Modeling of Gas-Fluidized Beds. In Marin, G. B., editor, *Computational Fluid Dynamics*, volume 31 of *Advances in Chemical Engineering*, pages 65–149. Academic Press.

van Leeuwen, I. M. M., Mirams, G. R., Walter, A., Fletcher, A., Murray, P., Osborne, J., Varma, S., Young, S. J., Cooper, J., Doyle, B., Pitt-Francis, J., Momtahan, L., Pathmanathan, P., Whiteley, J. P., Chapman, S. J., Gavaghan, D. J., Jensen, O. E., King, J. R., Maini, P. K., Waters, S. L., and Byrne, H. M. (2009). An integrative computational model for intestinal tissue renewal. *Cell Proliferation*, 42(5):617–636.

Vass, M., Shaffer, C., Ramakrishnan, N., Watson, L., and Tyson, J. (2006). The JigCell Model Builder: A Spreadsheet Interface for Creating Biochemical Reaction Network Models. *IEEE/ACM Transactions on Computational Biology and Bioinformatics*, 3(2):155–164.

Villasana, M. and Radunskaya, A. (2003). A delay differential equation model for tumor growth. *Journal of Mathematical Biology*, 47(3):270–294.

von Bertalanffy, L. (1950). An Outline of General System Theory. *The British Journal for the Philosophy of Science*, 1(2):134–165.

von Bertalanffy, L. (1968). *General System Theory: Foundations, Development, Applications*. George Braziller.

Walker, D. C., Southgate, J., Hill, G., Holcombe, M., Hose, D. R., Wood, S. M., Neil, S. M., and Smallwood, R. H. (2004). The epitheliome: agent-based modelling of the social behaviour of cells. *Biosystems*, 76(1-3):89–100.

Wallace, J. C. (1987). The Control and Transformation Metric: Toward the Measurement of Simulation Model Complexity. In Thesen, A., Grant, H., and Kelton, W. D., editors, *Proceedings of the 1987 Winter Simulation Conference*, pages 597–603.

Wang, J. (1998). *Timed Petri Nets: Theory and Application*, volume 9 of *Kluwer International Series on Discrete Event Dynamic Systems*. Kluwer Academic Publishers.

Weaver, W. (1948). Science and complexity. *American Scientist*, 36(4):536–544.

Webster, J. R. (1979). Hierarchical Organization of Ecosystems. In Halfon, E., editor, *Theoretical Systems Ecology: Advances and Case Studies*, chapter 5, pages 119–129. Academic Press.

Weiss, P. A. (1969). The living system: Determinism stratified. In Koestler, A. and Smythies, J. R., editors, *Beyond reductionism: New perspectives in the life sciences*, pages 3–55. Hutchinson.

Weiss, P. A. (1971). *Hierarchically Organized Systems in Theory and Practice*. Hafner Pub. Co.

Wiener, N. (1948). *Cybernetics: Or, Control and Communication in the Animal and the Machine*, volume 1053 of *Actualités Scientifiques et Industrielles*. The Technology Press, John Wiley & Sons.

Wilkinson, D. J. (2006). *Stochastic Modelling for Systems Biology.* Taylor & Francis.

Wilkinson, H. A., Fitzgerald, K., and Greenwald, I. (1994). Reciprocal Changes in Expression of the Receptor lin-12 and Its Ligand lag-2 Prior to Commitment in a C. elegans Cell Fate Decision. *Cell,* 79(7):1187–1198.

Wittmann, D. M., Blöchl, F., Trümbach, D., Wurst, W., Prakash, N., and Theis, F. J. (2009). Spatial Analysis of Expression Patterns Predicts Genetic Interactions at the Mid-Hindbrain Boundary. *PLoS Computational Biology,* 5(11):e1000569.

Wolkenhauer, O. (2001). Systems biology: The reincarnation of systems theory applied in biology? *Briefings in Bioinformatics,* 2(3):258–270.

Yamada-Inagawa, T., Klar, A. J. S., and Dalgaard, J. Z. (2007). Schizosaccharomyces pombe Switches Mating Type by the Synthesis-Dependent Strand-Annealing Mechanism. *Genetics,* 177(1):255–265.

Zeigler, B. and Praehofer, H. (1990). Systems Theory Challenges in the Simulation of Variable Structure and Intelligent Systems. In Pichler, F. and Moreno-Diaz, R., editors, *Computer Aided Systems Theory – EUROCAST '89,* volume 410 of *Lecture Notes in Computer Science,* pages 41–51. Springer.

Zeigler, B. P. (1976). *Theory of Modelling and Simulation.* Wiley Interscience.

Zeigler, B. P. (1984). *Multifacetted Modelling and Discrete Event Simulation.* Academic Press.

Zeigler, B. P. (1987). Hierarchical, modular discrete-event modelling in an object-oriented environment. *SIMULATION,* 49(5):219–230.

Zeigler, B. P., Praehofer, H., and Kim, T. G. (2000). *Theory of Modeling and Simulation: Integrating Discrete Event and Continuous Complex Dynamic Systems.* Academic Press, second edition.

Zeigler, B. P. and Sarjoughian, H. S. (1999). Support for Hierarchical Modular Component-based Model Construction in DEVS/HLA. In *Simulation Interoperability Workshop, Spring SIW.*

Zhang, L., Athale, C. A., and Deisboeck, T. S. (2007). Development of a three-dimensional multiscale agent-based tumor model: Simulating gene-protein interaction profiles, cell phenotypes and multicellular patterns in brain cancer. *Journal of Theoretical Biology,* 244(1):96–107.

Zurcher, F. W. and Randell, B. (1968). Iterative Multi-Level Modeling – A Methodology for Computer System Design. In Morrell, A. J. H., editor, *Information Processing, Proceedings of IFIP Congress 68,* volume 2, pages 867–871.

Zylstra, U. (1992). Living things as hierarchically organized structures. *Synthese,* 91(1–2):111–133.